Contents

DESIGN AND TECHNOLOGY IN THE PRIMARY SCHOOL

The inclusion of technology among the National Curriculum foundation subjects is an exciting but at the same time daunting challenge for primary

Subjects in the Primary School
Series editor: Professor John Eggleston

English in the Primary School	*Tricia Evans*
Geography in the Primary School	*John Bale*
Science in the Primary School	*Yvonne Garson*
Mathematics in the Primary School	*Richard R. Skemp*
Art in the Primary School	*John Lancaster*

The present trend in English primary education is to have teachers who are subject specialists as well as general class teachers. The books in this series are intended to support this specialist approach. Other volumes cover Art, Geography, English, Mathematics and Science, and a volume on History is forthcoming.

DESIGN AND TECHNOLOGY IN THE PRIMARY SCHOOL

CASE STUDIES FOR TEACHERS

HIND MAKIYA AND
MARGARET ROGERS

London and New York

First published 1992
by Routledge
11 New Fetter Lane, London EC4P 4EE

Simultaneously published in the USA and Canada
by Routledge
29 West 35th Street, New York, NY 10001

Typeset by Witwell Ltd, Southport

Printed and bound in Great Britain by Richard Clay Ltd, Bungay, Suffolk

British Library Cataloguing in Publication Data
A catalogue record for this book is available from the British Library.

ISBN 0-415-08089-4
0-415-03240-7 pbk

Library of Congress Cataloging-in-Publication Data

Makitya, Hind, 1952-
Design and technology in the primary school/Hind Makiya and
Margaret Rogers.
p. cm.
Includes bibliographical references and index.
ISBN 0-415-08089-4. -- ISBN 0-415-03240-7 (pbk.)
1. Technology—Study and Teaching (Elementary) —Great Britain.
I. Rogers. Margaret, 1945- . II. Title.
T107.M34 1992
372.3′58′0941—dc20

91-48095
CIP

Plates

Figures

Acknowledgements

Our thanks go to: Keith Wheeldon, HMI, formerly Art and CDT Adviser for Haringey, for making all the work described possible and for supporting us in the process of writing this book; John Eggleston, Series Editor, for giving us the opportunity to record our experiences and thoughts in this form, and for his patience and help in seeing it through to the end; Helen Fletcher, Martin Grant and Jacky Plaster, who have given so freely of their time, thoughts and ideas; Margaret Shaw, Sue Toogood and Christine Watson for introducing Hind Makiya to the very best examples of good primary practice in their own classrooms; Jeannette Campbell and Peter Clarke, who have worked closely with Hind Makiya in realizing ideas and developing them further; Andy Lambert (IT Inspector in Haringey), Richard Martin (IT Advisory Teacher in Haringey) and Alan Vallis, for help in the production of some of the illustrations in the book; Alan Vallis and Joel Xhaard, for their patience and support throughout; and all teachers and headteachers, named and unnamed, who have worked with us in the last four years – this book would not be here without them.

Hind Makiya
Margaret Rogers

1

Design and technology in the primary curriculum

The incorporation of design and technology in the National Curriculum is one of the most exciting developments in primary education in recent years. Pupils thrive on the types of activities that are involved, and the impact of this involvement and excitement is quick to reach the home, as a parent commented when a school organized a 'Design and Technology Workshop Evening' for parents: 'Sharon never talks about what she does in school – now all I hear is Design and Technology this and Design and Technology that. Do they do anything else in school, and what is this Design and Technology?' In fact, primary schools have been involving their pupils in design and technology activities for a long time. However, there is a need to make these activities more explicit and to co-ordinate them so that they can be used more effectively to draw out different curricular areas. This factor, together with the excitement pupils feel through engaging in design and technology activities, leaves us in no doubt that the subject has to be placed more centrally in teachers' plans of work. One of the main aims of this book is to demonstrate how design and technology activity can be used to draw out the other areas of the curriculum.

The introduction of new tools, materials and areas of technology and the need for teachers to familiarize themselves with these whilst engaging in practical problem-solving activities are paramount. However, whilst pupils are excited by the work that is developing in schools, many teachers are worried about their own training and confidence to handle the skills and concepts they associate with the area. We hope this book will both provide teachers with a variety of suggestions for organizing the classroom, including different methods of introducing the work, and stimulate them to develop a wider range of possibilities than those outlined in the case-studies.

Before describing the work in this book, we feel it is important to raise a number of questions and issues that teachers need to consider when they are involved in work described as 'design and technological' in nature.

- What are design and technology activities?
- Where do design and technology activities fit into the primary curriculum and what do they offer pupils?
- Is there a place for teaching design and technology across the curriculum in the National Curriculum?
- What is the gender dimension of design and technology?
- Where is the cultural dimension of design and technology?

What are design and technology activities?

In the *Interim Report* of the National Curriculum Design and Technology Working Group (1988) it is argued that design and technology:

> Is always purposeful (i.e. developed in response to perceived needs or opportunities, as opposed to being undertaken for its own sake), takes place within a context of specific constraints (e.g. deadlines, cash limits, ergonomic and environmental requirements as opposed to Blue-sky research) and depends upon value judgments at almost every stage.
>
> (p. 4, para.1.11)

Design and technology activity is seen as the process of satisfying needs by solving practical problems that involve pupils in working in a variety of materials. These materials include clay, paint, wood, metal, plastics, fabrics, leather and many others.

In this book we concentrate on a narrower range of materials, processes and techniques, because teachers are concerned about how they can be used in classrooms and across the age ranges. We also include concepts, such as mechanical movements, electricity, energy and control, that many primary teachers have not been familiar with. All the projects described in the book were done by classroom teachers before the publication of any of the National Curriculum documents on design and technology. We believe, however, that the methods of working and the philosophy behind them are consistent with all the attainment targets and statutory orders of the design and technology profile component (i.e. profile component 1: design and technology. See Figure 17).

Where do design and technology activities fit into the primary curriculum and what do they offer pupils?

Design and technology is regarded as a foundation subject in the National Curriculum. Its inclusion into every primary classroom is now statutory. The technology document has been published, indicating attainment targets and appropriate programmes of study for design and technology at different levels. The emphasis of the report is on the processes particular to design and technology activities: namely the design processes which involve solving practical problems within contexts that have meaning and relevance to pupils, and according to their abilities and experiences. The phrase 'problem solving' not only has several meanings, it is becoming an over-used expression and, in many instances, it is no longer clear what type of problem solving one refers to. In this book we hope to distinguish between three types of problem solving:

Puzzle problems

These refer to problems where teacher and children are aware that a specific answer is required and the teacher looks at ways to arrive at the correct answers. These are often referred to as 'closed problems' and the process is often immaterial. Examples can be seen in 'sums' in arithmetic, or punctuation exercises in English, or such questions as 'what is a square?', 'what is the name of this tool?' These problems are 'fact'-based and are used by teachers to check the understanding of pupils in a particular area. Many tests, such as comprehension tests and mathematics tests, are examples of these. The onus is on the pupil to answer the questions correctly. The process by which the pupil has come up with an answer is often not important.

Technical and investigative problems

In this situation, although there may be a series of solutions, the criteria for judging these are objective enough for participants to realize that some solutions are better than others. However, the process by which pupils have solved the problems is paramount. These problems tend to be subject-specific, i.e. mathematical investigations, science investigation including the design of fair tests, and technical problems. One example is the investigation into which materials float or sink (chapter 5); another, calculating the amount of wood used in a construction (chapters 2 and 7).

The processes by which these problems are 'solved' are very important. Science method, which accounts for 50 per cent of the assessment at the age of eleven (Science AT 1), is concerned with the skills of hypothesis, carry out tests and monitoring and recording methods leading to a particular evaluative statement–skills which are essential to the development of scientific rigour, as well as the fairness of the tests.

A mathematical investigation, such as finding the area of the school playground, may require pupils to discuss this in small groups, devise a method and then apply the method to find the result, either individually or in partnership groups. The decisions about how to proceed with the measurements are of greatest interest in analysing the pupils' understanding of the particular mathematical concepts. The actual measuring and following recording enables the teacher to monitor application skills of measurement, pupils' use of computation skills, estimation and concepts of accuracy.

Building the fastest moving buggy to cross a particular distance is an example of a technical problem, where the criteria are also specific and it is possible to judge and evaluate the most successful solution.

For the teacher, it is possible when devising these problems to plan carefully a series of activities that enable particular attainment targets to be addressed across the different areas of the curriculum.

Open-ended problems

These are also referred to as 'design problems'. They are more complex. The criteria for evaluating the work involve objective criteria (such as: how does it serve a particular function? how well is it made?), as well as subjective criteria (how does it look? is it appropriate for that particular need? how ingenious or 'appropriate' is the solution?) It opens up a debate about what is 'good'

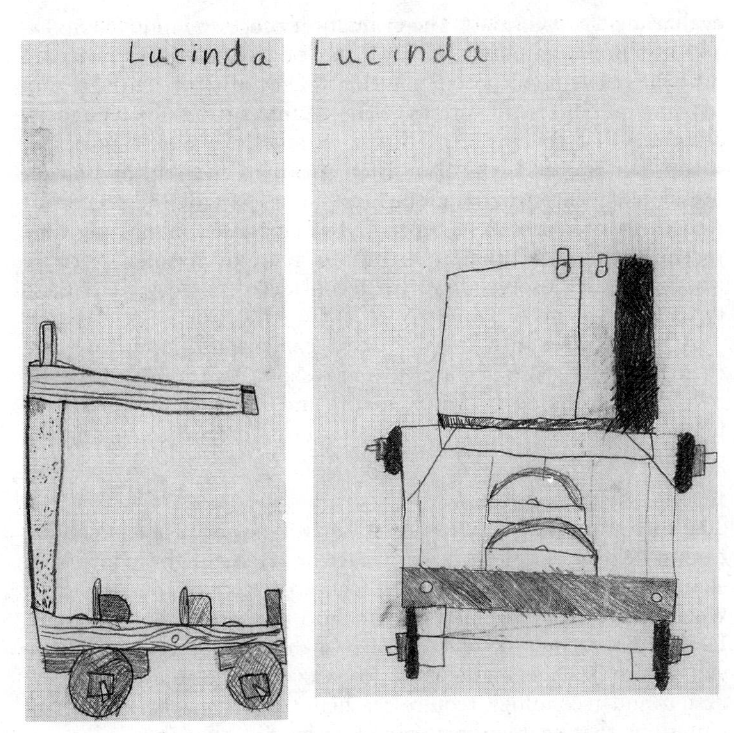

can you make a
buggy in Wood like
mine Lucinda

Plate 1 The buggy

design and what is 'bad' design, about who makes these value judgements and what these values are based on.

The pupils will be involved in a process of compromise between these subjective and objective criteria. The process of compromise and decision making becomes important in discussing and evaluating the solutions. The evaluation process is more complex, and it is not always clear which is the 'best' solution. Open-ended problems have many 'good' solutions. They also reflect the reality of many daily problems, as well as those faced by designers, architects and economists. They raise issues of value, choice, and social, cultural and ethical decision making. They are particulary useful in raising debate amongst pupils as part of the process of resolving differences in perception. The emphases on making value judgements and helping pupils build a sense of personal aesthetic and design philosophy have been considered as central to these types of problems.

A problem set in a social context can inspire pupils and give them a greater sense of purpose and reason. As a result, the work involving technology, science, maths and language that has been planned and linked to the project becomes integral to solving the problem.

The differentiation between these three types of problem solving cannot be seen as rigid. They are described to highlight many issues that are central to design and technology activity and to show where design and technology fits into the primary curriculum. Design and technology is more than a distinct subject area in the curriculum (such as mathematics or science). The requirements for design and technology require teachers to involve their pupils in situations that enable them to engage in the following:

- solving problems with some form of meaning to the pupils
- responding to a variety of needs
- thinking divergently and convergently
- making use of resources to solve the problems
- manipulating and using a variety of materials and media
- exchanging ideas
- evaluating ideas and products

The responsibility for teachers is to devise activities and make opportunities for the pupils to develop skills of divergent thinking as well as convergent thinking, and to help pupils develop appropriate communication skills; for example three-dimensional drawing, speaking to a group or responding to ideas. These are not innate skills, but ones that require consideration and planning. The pupils' love of the work is clearly linked to the practical nature of it, and the opportunities pupils are given to think and solve problems with some purpose behind them. The purpose and setting of the problems can never be underestimated. They provide pupils with the meaning and incentive for the work. They also have cultural and gender implications.

In primary schools, these problems are set in the context of topic work. The teacher or pupil, according to circumstances, focuses on an aspect of the topic and sets the scenario for a design and technology activity or problem. This scenario provides the 'context' for the problem. The planning, resourcing and build-up of this context is essential for the following success of the work. It also enables the teacher to build on pupils' experiences and widen them. The following are some examples of contexts picked out by some teachers:

Example 1

'A number of small creatures have lost their homes as a result of the storms. They need to find new ones that will be comfortable and safe from foxes and squirrels that could attack them. Many tree-stumps remain after the storm. Use these stumps to devise a home for these tiny creatures. You must work out a way for them to get up into their homes, as well as coming down, and being safe from predators.'

The pupils had to build part trees with papier mâché techniques. The various results reflected the different concerns of the pupils. There were a number of ingenious and very different solutions, appropriate to the problem and the particular creature the pupils had developed (this was part of the problem to solve). Through this project, the pupils had produced environments (the home) as well as designing systems (to get up and down).

Example 2

Making pop-up cards using different paper engineering methods provided the focus for design and technology activity within the topic of festivals. Pupils were asked to bring in the cards, purchased in high street shops, that they received for various festivals. The lack of appropriate cards for many different festivals and the mono-cultural nature of the cards brought in were highlighted as issues for discussion. Pupils were asked to choose a festival or celebratory event and design and make an appropriate card for a chosen friend.

In National Curriculum terms, the context was the community as well as business and industry. The pupils devised different systems for 2-D movements and produced artefacts in the form of finished cards.

Example 3

The topic was hats, and the teacher wished to introduce a further theme to the hats that pupils would be designing and making. She decided to choose 'movement'. Pupils were asked to design a hat that would incorporate either a hand-operated mechanical movement, or an electrical or pneumatic movement.

The examples listed above are also good examples of open-ended problems. Teachers may be hesitant about embarking on these open-ended problems, because of the fear that they may encounter questions from the pupils that they cannot answer or resolve. In our experience teachers have engaged in more complex problem solving as their confidence grows with the use of tools and knowledge of materials and processes. As there are no 'wrong' answers but many possible solutions, the challenge is as exciting for the teacher as it is for the pupils who are developing and refining their work. One teacher recently described her situation by saying 'I have developed the confidence to say to the children, "I don't know. Let us go and find out." She added, "I also know more about where to go to find out."

The skills, concepts and attitudes that are developed through an involvement in open-ended problem-solving activities cannot be underestimated. The list of elements of learning central to open-ended problem solving, shown in Figure 1, is offered as an aid to teachers considering these activities. It sets out the skills, concepts and knowledge that are required through design and technology activities with the attitudes developed as a result of trying out solutions – for example, perserverance and self-confidence. The chart is by no means exhaustive, but it has been used by teachers to plan for progression in skills and concepts in order to work with open-ended problems and is based on the HMI framework for the curriculum 5–16.

Elements of Learning / Areas of Learning	1 Skills	2 Concepts	3 Attitudes	4 Knowledge
Aesthetic & Creative				
Human & Social				
Linguistic & Literary				
Mathematical				
Moral				
Physical				
Scientific				
Spiritual				
Technological				

Figure 1 A framework for the curriculum

Is there a place for teaching design and technology across the curriculum in the National Curriculum?

At the time of writing this book the statutory orders for English, mathematics, science, history, geography and art demand that teachers address thirty-two attainment targets for Key Stage 1 (age 3–7) pupils, and thirty-three for the Key Stage 2 (age 7–11). These requirements do not take account of the other areas of learning that teachers are also required to plan for, including the human and social, and the moral. Teachers will still have to provide for a broad and balanced curriculum.

The management of such a wide set of demands to form a variety of stimulating, exciting and meaningful programmes of work requires careful planning and organization. The topic or thematic approach has been used as a means of showing how knowledge can be interrelated. We argue that this approach is even more important now, so that teachers can plan to address the wide scope of attainment targets by the careful choice of activities that can bring together seemingly diverse areas of knowledge. However, this planning process has to be structured and recorded in a manner that will enable parents, headteachers and other members of staff to understand the scope covered, so that there is some form of monitoring progression and continuity in the school.

All the case-studies described in this book have been planned as part of a topic. Teachers have devised projects for pupils to design and make. Around the projects, they have devised a series of activities, linked to the work, and have used these as a means of addressing different curricular areas. In chapter 8, we will return to the issue of topic planning and offer teachers a model for planning work that may be used and adapted to suit their needs. We will show how teachers, through skilful intervention and the provision of appropriate starting points for the learning process, can use their considerable expertise in devising projects to introduce design and technology activity to their pupils.

What is the gender dimension of design and technology?

Gender becomes an issue every time the pupils pick up tools, switch on the computer or reach for the Lego, and at whatever stage the pupils are in their primary education. Design and technology in primary schools is providing both educationists in general and the design and technology community with an exciting challenge. Girls are proving their abilities and their love of the work. The primary workforce, with a large majority of women, is challenging the traditional stereotypes associated with design and technology and providing pupils, and girls in particular, with long-needed role models. Pupils, of both sexes, are using tools at an ever earlier age and engaging in technological activity that would formerly have been introduced to 13-, 14-, and 15-year-old pupils.

There is much research carried out about the different notions of technology and the gender differences between these. It is not our intention to discuss these here. However, it may be useful to raise some issues that are particularly relevant to the types of activities that teachers would consider in their classrooms and the possible gender implications of these.

First, access to tools, materials, construction kits and computers, as well as opportunities to explain projects and offer suggestions, are two very important issues for teachers to consider. Whilst traditional roles are being challenged in the classroom, pupils come to school with different past experiences and role models (they may or may not have Lego or tool kits at home). It is important that experiences are recorded and issues of access discussed and monitored. Most of the work described in this book has been carried out with small groups of pupils working together. Teachers have choices to make when grouping pupils. It may be possible to have girls working together, so that there is less likelihood of 'tool hogging'. However, experience has also shown that it is important to change groups so that girls work with boys and actually confront any issues arising from the sharing of tools or lack of experience of working together.

Second, the choice of project and types of problems we ask girls and boys to engage in are very important. These can be used to challenge stereotypes and highlight gender issues. Their careful choice can also broaden pupils' perceptions of what is technology. Teachers can ask themselves a series of questions at the start of their classroom planning process with regard to the types of situations they make available to their pupils, and what they hope to draw out from them:

1 Why am I choosing this particular problem/investigation? Does it raise any particular gender bias? Are there any alternatives, equally suitable and more widely appealing? Example: it is possible to engage pupils in finding out about wheeled vehicles without having to make cars. The examples of the wheeled dragons in chapter 3, the carnival floats, machinery or roundabouts are some of the alternatives.
2 Can I use the work to tackle positively issues of gender? Example: puppet plays provide pupils with the opportunity to act out their attitudes and experiences and confront stereotypical images often raised in fairy tales.
3 Can I use the work with the tools to raise questions, such as, who uses tools? what do we mean by tools?
4 How can the teacher discuss what pupils associate with 'technology' and broaden these notions? Example: what would constructing a dress pattern and making the garment be called?

Third, research has shown that girls are more interested in tackling open-ended problems where there are value judgments to be made and different needs to be responded to. (Grant and Harding 1984). Projects such as the fastest/slowest buggy leave many pupils at the back of the classroom unenthused by the seemingly arbitrary challenge. Once again, the choice of work and an identified sense of purpose to the project are very important considerations for the teacher.

Finally, primary women teachers also have a duty and responsibility to ensure that they participate fully in this aspect of the curriculum and not leave the work to keen male colleagues. There are many national and local INSET courses available across the LEAs, and women teachers need to insist on attending these so that they are able to build the confidence of all colleagues within the school. The importance of avoiding stereotyping and tackling sexist prejudices and practices cannot be overstated.

Where is the cultural dimension of design and technology?

The need to involve more girls in all aspects of technological education has been receiving greater recognition on a national scale. It is also of paramount importance that technological education at primary and secondary levels should address the concerns of all pupils and that this education reflects the multi-cultural contributions to design and technological achievement across the world.

The issues we address above should be regarded as important in terms of race as they are in terms of gender, for it may be feasible to develop work that is anti-sexist in nature whilst being racist, if it is based on a biased cultural perspective. In the National Curriculum Design and Technology Working Group's Proposals, *Design and Technology for ages 5 to 16* it is stated:

> It is important that teachers take a positive approach to a mixed range of cultural backgrounds in their pupils, rather than an approach which concentrates on the problems that some pupils may have in coping with, for example, the language of design and technology. The variety of cultural backgrounds of pupils can broaden the insight they all have into the range of appropriate, alternative solutions to perceived problems. There are rich opportunities here to

> demonstrate that no one culture has a monopoly of achievements in design and technology.
>
> (p. 7, para. 1.46)

Similar questions can be asked about the relevance of the work, and whether it can be used as a focus for challenging stereotypes. Other questions that are particularly relevant in this instance would include: Who decides what is good design? How can we develop in pupils a sense of aesthetic awareness without imposing our own values of 'good design'?

Technological development and innovation has a major impact on all the world today. This innovation can change a cultural identity, and such changes are open to debate with regard to their positive and negative effects. It is important to engage pupils in such issues through careful planning and preparation. For example, recent debates about modern architecture in Britain can be paralleled in the Middle East: how can we develop a changing architecture that is based on some understanding of the historical background, geographic circumstances and cultural heritage of a particular area?

Much of our modern art and design stems from African and Asian inspiration. The source of designers' influences need to be highlighted to pupils, since they widen their perspectives. The wood and highly polished stainless steel sculptures of Brancusi are directly influenced by early African tribal sculpture. It is possible to draw many examples of inspiration from many other countries as further examples. The history of technological innovation also shows the various technological achievements from other cultures. Teachers should question whether they are including these aspects in their work.

Since the design and technology profile component has such a strong emphasis on evaluation (AT 4) and identifying and satisfying human need (AT 1), and because value judgements are very strong aspects of both these areas, it is necessary for teachers to consider how pupils build these value judgements and the factors that influence these. Gender and culture are two of these factors. Culture can also include peer culture, and social and economic cultures apart from historic and geographic factors.

The potential of design and technology in the primary curriculum

Several issues have been raised in this introductory chapter in the hope of exciting the reader with the potential and challenge of design and technology and in particular its potential in the primary curriculum. We also hope they show how much further work needs to be developed in this area. One of the main objectives of this book will be to demonstrate, through examples of project work, that design and technology activities are central to the primary curriculum. We believe that they should be the focus of the work in the classroom. This will be shown by examples of that work.

All the work has been done by classroom teachers with minimal external assistance in their classrooms. Many teachers have attended in-service training organized by the local education authority, aimed at developing their confidence to engage in handling a wide range of materials, techniques and a few simple tools.

Case-studies are presented indicating the topic and age of pupils. Experience has shown that pupils can engage in constructional work at a very early age. It is possible to use any of the ideas described and develop them for the appropriate year and Key Stage. All the projects discussed in this book use design and technology activities as a focus for the topic chosen. This has proven very successful, from both pupils' and teachers' points of view.

Chapters 2–7 are set out as a series of case-studies, showing work with pupils in Reception, Key Stage 1 and Key Stage 2, how they are introduced to tools and ways of organizing the classroom. The projects can be developed and adapted to different ages, depending on the objectives of the teacher and the particular circumstances. The levels of achievement of the Reception class can be seen as inspiration for planning progression through the infant years. We hope they also raise the question, 'If this is what they are able to do at the age of 4–5, what will they be able to do by the age of 11 or 12 given continuity and progression?' The cross-curricular nature of design-and-make activities is also drawn out through a series of case-studies in chapters 4 and 5, where projects have been planned to engage pupils in mathematical and scientific skills and concepts.

The recent rate of change and curriculum development in schools has meant teachers are no longer able to concentrate on their own classes, but are also part of a wider team. In-service courses, curriculum innovation, and school planning and change become the focus for development. School development and topic planning are looked at in chapters 7 and 8.

In clarifying what design and technology activities are, we have described the three types of problem solving which also demonstrate links with science and mathematics (technical and investigative problems) and make possible the bringing together of science, technology, mathematics and language work to develop pupils' attitudes and values in open-ended problem solving. We favour the latter because of the opportunities offered to tackle gender and cultural issues, which we see as central to all the projects described in subsequent chapters. We begin and end this book with topic planning. The National Curriculum and specific attainment targets for design and technology are referred to in the relevant chapters.The Attainment Targets for the Design and Technology Capability Profile Component are:

- AT1: identifying needs and opportunities
- AT2: generating a design proposal
- AT3: planning and making
- AT4: evaluating

whilst Attainment Target 5: information technology, is a cross-curricular activity and is specifically referred to in chapter 6.

2

The early years

Pupils of nursery and lower infant ages come to school with expertise in solving problems, manipulating found objects, finding meanings in their surroundings and making choices, and with massively enquiring minds. The role of the teacher is to use these design and technology capabilities and build upon them, as well as introduce them to a variety of new techniques and skills that will extend and widen these experiences.

In this chapter, we focus on examples of work with Key Stage 1 pupils. We describe a series of projects undertaken by a Reception class to illustrate ways in which design and technology activities can be organized with the younger age-ranges and within the context of the topic 'Ourselves'. Through a series of planned activities, pupils are introduced to tools and different working techniques. Readers will be able to see how this project work supports and is supported by other areas of the curriculum, for example language, mathematics and drama.

One of the problems that faced the Design and Technology Working Group was the lack of sufficient recorded evidence of the capabilities of 7-year-olds who had been engaged in design and technology activities in a systematic manner that had incorporated progression, and who had started at the age of four or five.

We have therefore chosen to focus on the work of a Reception class (4–5-year-olds). Whilst the statutory orders of the National Curriculum would not apply to the majority of the class for their first two terms at school, we have shown the potential of the work in terms of attainment targets that can be addressed through these activities. The work also shows how it is possible to plan for a series of experiences that incorporate progression and diversity of experience.

Topic: 'Ourselves'

School: Rokesly Infant School
Teachers: Ms Christine Watson and Ms Annie Stotesbury
Age-range: Reception (4–5-year-olds)

Aims and objectives

The class teacher had decided to work on the topic 'Ourselves' throughout the year, in order to record the children's first year at school. By the end of the year each child would leave with a book about themselves, with examples of work and photographs of themselves at school.

This description concentrates on the practical parts of the overall classwork and on a choice of three projects in particular. These have been chosen because they show how the teacher had planned for a series of objectives to be met through the work:

- introducing pupils to a series of tools and developing their confidence in using them (paint-brushes, saws, vices, abraders, drills, scissors)
- acquainting pupils with different materials, reflecting on where they are found and what they are called (waste materials, resistent materials)
- introducing a number of techniques for working the tools and materials, including joining and gluing methods, as well as simple finishing techniques
- giving the pupils opportunities to plan, make choices and implement them
- using the activities above as a means of engaging pupils in practical science, mathematics, drama and art activities
- using the work as a means for making class reading books

Project 1: making masks

The class had been introduced to the paints and to simple techniques of colour mixing. They then proceeded to draw large pictures of each other, looking at different parts of the face and the relationship between these. The pupils were encouraged to mix the paints to try and achieve their own skin colour (AT 1).

The next step, in terms of developing manipulative skills and

Plate 2 A helping hand builds confidence in using tools

modelling abilities, was to make heads, using paper bags filled with newspaper and some simple 'junk modelling' materials, such as cardboard and plastic containers, to make heads. (AT 3).

The whole class was introduced to the tools; for example vice, saw, hand-drill, abrader (see the Appendix for more details). Methods of handling and carrying the tools were established to ensure safety. The pupils were shown a variety of examples of what they could make, the object being to introduce them to the range of possibilities without narrowing their vision or hindering their own ideas (AT 1, AT 2).

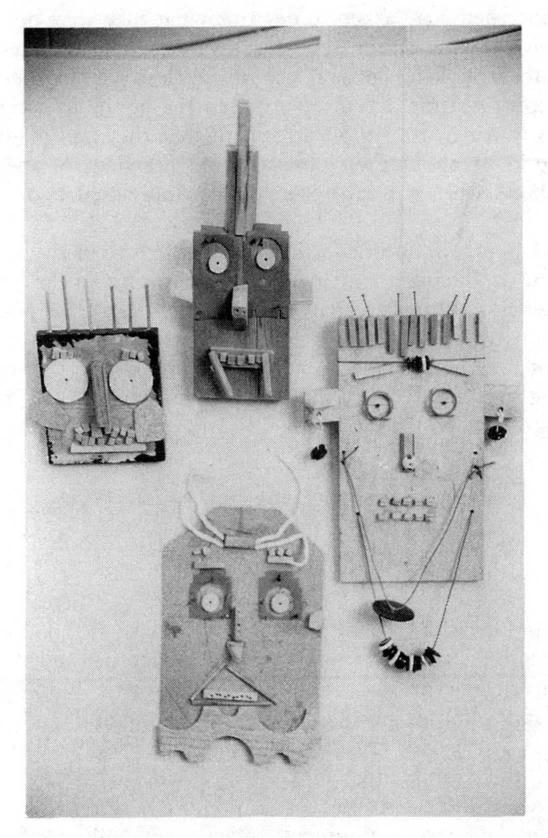

Plate 3 Masks made by Reception pupils

The teacher took groups of six to make their masks in a morning, working individually so that all had access to a saw and a vice. Other tools were shared. Initially the teacher helped them with the drilling, but as work and confidence developed pupils helped each other. They each chose a backing (mostly hardboard pieces brought from a local scrap project) and proceeded to work out how they would use their cut pieces to make up the parts of the face (AT 3); see Plate 3).

In the mean time a number of photographs were taken of the pupils at work. These were mounted on card, and the class and pupil concerned were asked to describe what they were doing. The teacher wrote this down. These cards were then laminated and made into a book for general use in the classroom. Other simple introductory writing activities involved the pupils in copying the teacher's writing of a simple statement that they had made about their work, or about their observational drawings of the masks. These sheets were put into each pupil's individual book (AT 3, AT 4).

Year 1 pupils in another school were able further to extend the project by producing a 'cutting list'. This is a list in the form of a chart, which indicates the part or component (e.g. 'the nose'), the material used and its measurements. Cutting lists are normally made out before making an object to help in costing the work and clarifying sizes and types of materials. In primary schools, they can be used to help pupils in the classification of materials as well as measuring. Key Stage 2 pupils could also carry out costing exercises.

Through this initial collage project, the teacher had introduced the class and each pupil to the safe use of the tools. They had constructed two- and three-dimensional masks using different materials. They had to plan their simple faces, thereby making choices and decisions. They had to explain why they had used the different pieces, and what they were (to the teacher and the class). They had engaged in some simple recording of the work, written, drawn and photographed. Pupils were designing and making artefacts. They had reflected on the processes and they had followed and evaluated their work in a series of 'show and tell' discussions in the classroom.

All the language attainment targets were addressed in this project and through the work of the pupils. They had been involved in making a class book and building up their own personal book. They were engaged in simple writing exercises, to enable them to draw simple letter shapes and convey their expressed ideas. They had to work together and individually. They had to show and explain in their small groups what they had done.

In terms of science, they were starting to address the processes of observation, planning and recording. Other science attainment targets that were being addressed were highlighted through the discussion on colour pigmentation, the parts of the face, the use and naming of the different materials.

In terms of mathematics, they were introduced to the concepts of two and three dimensions, as they worked with the paper and paints and then made three-dimensional masks. When they were cutting their pieces, they were encouraged to cut long and short pieces, as well as pieces which were 'the same as' others. Simple counting exercises were devised, based on what they had made, and the classification of materials into different categories was used to highlight different mathematical areas of shape, number and measurement.

Project 2: making puppets

Now that the pupils had experienced the use of tools, the aim of the second project was to introduce a greater element of planning and forecasting. The theme of the project was puppets. A discussion about the parts of the body and the joints in the body helped the pupils to think about knee, elbow and ankle joints. Once again they were shown a variety of simple wooden puppets with different jointing methods: for example, twisted wire, threading, screw eyes, staples, card and material hinges. (AT 1, AT 4).

The children used simple shapes found in the mathematics corner (Logi Blocks) to draw the parts of their puppets. The main geometric shapes were highlighted and the pupils drew around these on large sheets of paper. Then they looked for pieces of wood that would be suitable to make their shapes. They had to saw and shape these to match their drawn shapes. They could then choose the most suitable method for joining their pieces to make the puppets. The puppets were finally strung up on two crossbars, and the children decorated these with a wide variety of materials, applying paints and simple decorative finishes (AT 2, AT 3).

The pupils worked on their puppets individually, but were encouraged to ask each other questions first. The project took the equivalent of one and a half school days to complete, with groups of six children working together at a time (AT 4).

The teacher's main concerns in this project, were to observe how the pupils used the tools, given their previous experience, and how they attempted to mark out pieces that were the same as those they had drawn; to look at the degree of accuracy of each pupil's work, whilst talking to them about their work and observing their use of language in terms of knowledge of the names of materials, tools and the shapes they were cutting. Other areas not included previously were joining techniques for the different joints of the body, in order to achieve some form of movement and control.

Familiarizing the children with basic geometric shapes, as well as trying to reproduce them, were excellent exercises in shape recognition as well as further development work on the comparative measurements built up from the previous project. Pupils were using and applying mathematics. Number work was encouraged through the questions pupils were asked; for example 'how many pieces of plastic did you use?' or 'how many circle shapes do you have in your puppet?'

The pupils were also encouraged to estimate how many pieces were used for the teeth in a particular face, and then to check their results by counting. The different materials used and the classification of these was a theme throughout the year, and this involved mathematical concepts of sorting, as well as scientific concepts of understanding and classifying materials and finding out about their properties. In addition to this, the pupils were extending their knowledge of the different parts of the body and hypothesizing on how they are connected.

Puppets can be used in a variety of language situations to encourage the pupils to make their own plays. In this instance the puppet design and making activities served as a focus for the drama work and character building that followed and made use of the pupils' puppets.

Project 3: where we live

This project describes one of the other activities carried out by the same class further on in the year. The class were introduced to a simple electricity kit (Middlesex Polytechnic kit). This kit can be used with a group of pupils (working in twos) to enable them to explore the principles of circuits. Success is easily achieved, whilst children are allowed to investigate and develop the circuits, raise hypotheses and test them out, sharing ideas with each other. The kit is easy to use, and the pupils need not worry about connections. They can try out their ideas and get immediate results. Teachers have used it successfully to introduce the concepts of an electrical circuit, a power source, connectors, switches, conductive and non-conductive materials.

It was important to draw out the scientific knowledge specific to electrical circuits, as the pupils used the kit. Using this method means a sharing of information with others. In this case, the class were divided into pairs to work together, and a rota was set up for access to the electricity kit. Sharing and discussing the work and using terminology such as 'circuit' and 'battery' were activities that were highlighted during group discussion time, when each group reported back. They were asked to make predictions before testing and then find out if their predictions were right.

Alongside this work, pupils were also involved in mapping work, finding out where everyone in the classroom lived and actually visiting each child's house. A picture was taken of each child in front of their house. This project was undertaken at a time when the children were visiting another school, so this school was also involved (AT 1, AT 5).

A large map (8 × 10ft) was drawn, painted on paper and then stuck together. Each street was named and each child's house located on it with a bulb to light up the position of their house (see Plate 4). The pictures were attached to the wires, so that each person would be identified (AT 3).

The electricity kit on its own is not enough. Scientific knowledge is not enough. The 'hypothesize, test, experiment and evaluate' method is not enough. Children are interested in the applications of their knowledge, and this is the great potential of such work. In this instance, they had to transfer the knowledge they had learnt about electricity from the kit to small bulbs, wires and batteries to make their own circuit to place on the map. In this project, the whole class was involved in working towards a joint effort. The map belonged to the whole class. Projects like these help to develop feelings of communal responsibility and ownership.

Summary of the three projects

It is important for the children to have shared experiences (i.e. all of them have access to the same type of tools, the electricity kit, or the Technical Lego). However, they do not need to produce the same end results (different types of puppets will dictate different jointing systems). Through the processes of discussion and pupils' evaluation of their work, ideas and different solutions can be shared with others, thereby giving the class access to a wider range

Plate 4 Map made by Reception pupils showing each child's home

of experience (AT 4). Furthermore, this work can be introduced at any level, with differing results in terms of sophistication. This particular point is raised again in the 'transport problem' topic in chapter 5.

The three projects chosen were used to describe how a particular teacher introduced work to a Reception class. However, this case-study has also been used because it highlights a number of issues:

1 The teacher's initial aims and objectives, in terms of introducing the tools, materials and techniques of working, and develop-

ing pupil's competence in handling these, are all possible with pupils of a very early age.

2 We have shown how these projects can be used to satisfy the four attainment targets in the design and technology profile component. We have also indicated those particular skills and areas listed in the detailed programmes of study.
3 The importance of linking these activities to the overall topic, so as to draw out as many other curricular areas as possible. The design and technology activities and projects can be used as a basis for lively pupil-centred language, science and mathematical activities. However, these activities need to be planned carefully if they are to address specific attainment targets in other areas. In this instance, the teacher has carefully located what she wishes to draw out of the project.
4 The teacher has very carefully looked at the issue of progression. Progression in these projects can be mapped in terms of:

- the use of tools
- expanding the range of materials
- planning skills
- the type and complexity of the projects chosen, which included two- and three-dimensional work.
- using the projects to extend the number of subject areas involved following the work
- organizing and planning activities that develop individual work and group work, including pairs of pupils, small teams and the whole class. The development of group work is done gradually and by encouraging pupils to discuss their constructional techniques with each other

In this respect, pupils slowly develop an understanding of a particular concept (e.g. electrical circuits, mapping, shapes, joining methods) by continuously engaging in activities that require them to apply skills learnt and 'tinker' with the concept involved.

Other infant work

There are many more ideas that can combine different materials and areas of the curriculum. In the same school, with another class of the same age-range, every pupil designed and made a symmetrical necklace, using various materials and shapes. This project was very successful, because it combined aspects of planning and decision making, as well as the task of making a second piece 'the same as' the first piece. Comparative measurement became a very real problem. The pupils had to thread their work. They were also involved in looking at symmetrical and asymmetrical patterns in general.

It is also possible for children to transfer from modelling with the use of construction kits to modelling with found materials from a very early age. Kits enable them quickly to build, test and modify without having to commit themselves to lengthy making processes. When success is assumed, appropriate materials such as wood, card and plastic waste can be used to construct a permanent model.

In another school, older pupils (Years 1 and 2) used their classroom construction kits (Basic Lego, Technical Lego, Stickle Bricks, Reo Click and others) to design and build simple moving toys. They were asked to work in pairs and jointly decide what they would like to make and construct it.

Pupils enjoy making models from the kits they are accustomed to. They are confident in handling these and building models to fulfil various functions. The kits are easy to use and ideas can be tested out quickly. However, it is also important that the pupils are able to translate these ideas into their own products, so that they encounter the problems of adapting, constructing and manipulating materials. When they had completed their toys, they were asked to draw these in detail, so that they could develop their three-dimensional drawing skills. (For notes on the use of kits and Lego in particular, see the Appendix).

The pupils were asked to discuss what they thought about their models. Since another group would be doing the same work, the problem of breaking up these toys was discussed. This provided the pupils with the incentive to realize their ideas in more permanent forms. The next step was to ask them to consider how they would make their models using timber, dowel, wooden wheels, corks and various other materials.

They had been shown how to handle the tools. It was amazing to see how easily they translated the work from the kit to other materials. As other pupils saw this happening, the next models were even better. At one stage, the teacher could see one of the pupils using gears, and she advised the pupils that these might be too difficult to construct in wood, suggesting pulleys as alternatives. However, the pupil insisted, and proceeded to saw out teeth for the gears from the wooden wheels. The results were excellent and the gears meshed.

In this project the kits were used as a means of looking at how things moved and how they could also be controlled, as well as to generate ideas quickly and communicate these in simple forms. The pupils were building structures using an established kit system. Whilst the kits are good for drawing and planning, nothing can replace the satisfaction pupils gain from making their own individual model, which they can show to peers and parents. The need to break down the toys created a need for actually making the toys in other materials (AT 1). Such projects engage pupils in all the attainment targets particular to design and technology, and address many aspects of the programmes of study as well.

The projects described in this chapter can be and have been used by teachers in all the other years of Key Stages 1 and 2. In each case, the teacher has planned the work and the supporting activities to achieve particular ends and to be appropriate to the age and abilities of the pupils. The projects described in the next chapters are equally applicable to the younger age-range. Teachers continuously show how it is possible for ever younger pupils to engage in more complex work, provided it is planned and adapted to the particular age-range and placed in a context that has some relevance to the pupil.

3

Starting in the classroom

All three projects in this chapter describe work with first year junior classes (Key Stage 2: Year 3) in different schools. Our intention is not to describe the projects in any detail, but to look at possible classroom organizational methods for initiating work with tools. Since writing these case-studies, other teachers have used similar techniques with younger pupils, demonstrating the unknown limits and capabilities of pupils to do this work, given the motivation and love of practical activities as driving forces.

The first project is based on a farm topic. The second and third are based on a dragon project started at the time of the Chinese New Year. Each of these shows a different form of classroom organization:

1 Use of Year 6 pupils, who have previous experience as facilitators for Year 3 groups.
2 The teacher shows two groups of two pupils how to make a box. They are then each responsible for showing another pair of pupils the method.
3 The teacher makes use of another adult to help initiate the whole class to the use of tools at the same time.

The two dragon projects illustrate how it is possible to approach the same topic in different ways, also highlighting how a class project can raise valuable lessons on democratic decision making processes in the classroom. Both projects are based on building an articulated wheeled dragon. Such projects can be used to include a lot of work on wheels without having to resort to making cars or buggies, which are becoming popular 'technology' projects in many primary schools.

All three teachers have decided to use a prescriptive technique as a starting-point to the work. Reference is made to a frame or box construction often known as the 'Jinks' frame or box. This technique of gluing square-sectioned wood and reinforcing the joint with card triangles was described by Williams and Jinks

(1985). It was developed as the result of work in a primary school, where the frustrations children experienced with traditional methods of joining affected the potential of the work to develop.

A major feature of this method of construction is that it is possible to build light-framed structures, using simple tools and with little experience of their use. This gives confidence not only to the pupils but also to the teachers embarking on this work for the first time. There are other very successful methods of joining that have been developed by others in the field of primary design and technology. Garden canes and elastic bands have been recorded by Paul Shallcross (1985). Dowel joints are used by Jim Flood (1986). All these techniques are simple and have an in-built element of success. However, they are not the only methods available, and it is important to encourage pupils to use them appropriately, otherwise they become as restrictive as the traditional construction techniques they replaced, and they can end up determining the outcome of the project.

It is important that primary design and technology is not equated with a particular technique. None of the initiators of these techniques would wish that to be the case. However, it is possible to use such a technique as a means to introduce the tools in the classroom, as well as a means by which to engage in other curricular areas. Chapter 7 looks at how a whole school approached the issue of introducing the tools using the 'Jinks method'. This prescriptive method was used as a means of achieving a number of specific school objectives and as an interim measure to more open-ended problem solving.

Topic: 'Farms'

School: Rokesly Junior School
First Year Junior (Key Stage 2: Year 3)

In this school, some teachers had been on training courses on the use of tools, and the Year 5 and Year 6 classes had already engaged in exciting and challenging work (see chapters 5 and 6). There was considerable enthusiasm and excitement amongst other pupils about the work they had seen. However, there was a great deal of fear and lack of confidence among the teachers, who had not had the training and experience with the tools. A student teacher who

Plate 5 Boxes

had been to see the work during her teaching practice and had then started teaching in the school was keen to try out some work with her class. The problem was to find some form of support for introducing the work and carrying it through. The Year 6 pupils in the school were used as this support, providing technical expertise and ideas as appropriate. A number of Year 6 pupils volunteered, and their roles were discussed with them in detail. They were asked to write about their experiences with the Year 3 pupils as well. Each Year 6 pupil worked with one group of first years.

The topic was 'Farms', and the children had visited an inner-city farm and had looked at all the buildings (AT 1). The different buildings that would be built were decided upon and the class divided up into groups of three and four. Cowsheds, pigsties, sheep-pens, stables, farmhouse, barns and silos were the buildings chosen (and a farm tractor). The children drew full-size pictures of what their buildings would look like, and proceeded to build these using their drawings as a means of measuring and checking for accuracy (AT 2).

Apart from constructing the basic structure of the building, pupils had a number of other decisions to make, including those about the different roof structures and whether these would be removable or hinged so that they could look inside. A variety of materials were used to cover the buildings: corrugated card to simulate roofing sheets; veneer to simulate timber; woodchip wallpaper to simulate concrete rendering.

Children went outside the school building to look at brick patterns which were recorded. They then did some colour mixing to look at the different colours of the brickwork. These sheets of paper were then cut into individual bricks which were glued on to the main structure to give a more three-dimensional effect. Wood shavings were glued on to the external card that covered the building to achieve a rough rendered effect. Varnish was also applied to the buildings to give them a finished appearance. Some pupils made patterns on paper with potato block prints to simulate bricks (AT 3).

The students used an electricity kit to study circuits, achieving instant results which they then recorded through drawings and labelling. They discussed, as a class, what the requirements for an electrical circuit would be, before using their own components and applying them to their buildings (AT 1).

This work took place over half a term, but the pupils' enthusiasm held right to the end. It was also important that the children were given time actually to play with the farm they had built at the end of the year, in spite of some accidents to the buildings (AT 4).

The Year 6 pupils who helped also found the work with the Year 3 pupils interesting, since they were showing others techniques they had learned. It was important to remind them frequently that their roles were those of assisting and not actually doing the work. They were also asked to encourage the Year 3 pupils to think and plan their work both individually and as a team, in order to anticipate possible problems. This was done by asking them questions about what they were doing and the purposes behind it, as well as monitoring access to tools and a sharing of the work. This contact between Year 6 and Year 3 pupils helped to build a good relationship between the different age 'ends' of the school. It was interesting to reflect on how the Year 6 pupils learned the value of accuracy and precision from their previous work as they emphasized these skills to the Year 3 pupils.

It is possible to see from the project above how pupils had experienced aspects of each of the four attainment targets for design and technology profile component: by visiting the farm, deciding what buildings their own farm would include (AT 1/AT 4), planning what they would look like, and drawing different elevations of these (AT 2); by working to their drawings and making their buildings, encountering 'sub-problems' of construction, such as roofs, doors and windows (AT 3); investigating different materials in greater depth to attempt to simulate building materials, by having to make choices about which would be the most appropriate, and by using and playing with their farm, thereby evaluating it constantly through use (AT 4).

Topic: The two dragons

If teachers are exploring festivals and celebrations for their topic, January is the time when they often choose the Chinese New Year. This opens up a tremendous potential for colourful and eye-catching displays of both two- and three-dimensional work. This topic can be approached in a number of ways, involving discussions and exploratory work with other new year celebrations, such as the Nawruz New Year (where the first day of spring marks the beginning of the year) as well as January 1st celebrations that will have just passed.

The Chinese New Year can be used to explore exciting aesthetic and creative work through a study of the celebrations and dance that take place. It can be used to look at other aspects of Chinese culture, history and geography by focusing on China. China could be introduced as a country with both geographic and ethnic diversity, and, by comparing sizes in terms of area, population and cultural diversity with Britain, interesting activities can be devised to highlight issues particular to the teacher's plans, access to resources and possible visits, as well as the pupils' own experiences.

The following two accounts show how two teachers from different schools approached making a dragon for the new year celebrations. Through using the same topic we highlight the following two issues:

1 The importance of individual approaches to the same topic.

Although using the same starting-points for the topic and sharing similar aims and objectives, individual results can be very different. These differences highlight the potential for adapting ideas from other schools to suit particular teachers.

2 How a teacher's own areas of expertise and concerns are reflected in the outcome of the work.

The question that is consequently raised is: how can a school use an individual teacher's knowledge and expertise across the school? We look at this question in greater detail in chapters 7 and 8.

Before enlarging on how each teacher tackled the project, it is useful to look at some of the common objectives of the teachers involved.

Both teachers had attended in-service courses and were starting off in their classes for the first time. The pupils had no other role models in the school from which to understand what might be expected. The teachers had set themselves specific objectives for the project:

- Cope, personally, with starting this work in the classroom for the first time.
- Introduce every pupil to the use of tools and materials and to safe working procedures.
- Develop group work to enable the class to work co-operatively in groups of two.
- Involve the pupils in some elements of technical problem solving.
- Using the prescriptive construction technique as a means to develop sequential writing for producing instructions on how to make a box, or frame.
- Asking pupils to make detailed observational drawings of the tools they had used.
- Asking pupils to record their structures before covering them.
- To make a class book.

Topic: 'Dragons'

School: South Haringey Junior
Teacher: Ms Jeannette Chapman
(Key Stage 2: Year 3)

The class teacher had, at that time, recently been appointed with responsibility for overseeing science in the school and had taken on the task of studying the relationship between science and design

and technology activities. The whole school was engaged in design and technology curriculum development. The activities were seen as a focus for the topic work in every classroom. (The school is discussed in chapter 7). The initial focus of the work was the Chinese New Year. However, as the project developed, it moved towards dragons in different mythologies with the pupils reading dragon story-books and finally rewriting these.
Additional objectives particular to this project include:

- To focus on the particular science skills and processes that arise from this work.
- To make a study of the different materials used in this project and throughout the school, in order to develop a coherent school policy on materials, their use, storage and access.

During classroom discussion, the teacher explained to the pupils that they would be using tools for the first time to make simple box structures. These would be used as segments of a long articulated dragon.

After discussing what a box structure was called, the dragon was known as the 'Cuboid Segmented Articulated Dragon'. The teacher decided to organize the work in the following stages:

1 Pupils would be grouped in two to work together. Each partnership would make one cuboid structure, thinking about sizes before starting.
2 Each group would have to work out how many wheels were needed and at least two methods to attach wheels and axles to the structure
3 These wheeled structures would be placed in size order, and the two adjoining pairs of pupils would be asked to work out how they would join their two structures together.
4 The whole class, after testing and evaluating the results of the above, would decide a method for joining the pairs of boxes together.
5 During the above processes, they would cover their structures with card. Pupils were introduced to the notion of area, and were asked to work out how much card they had used.
6 A decision would be taken about the decoration of the linked dragon segments.
7 The dragon's head would need to be considered. The teacher had

anticipated using light bulbs for the eyes and wanted to incorporate simple pneumatic movement using syringes and tubing (see the Appendix).

The teacher explained these basic stages to the pupils, so that they would be aware of the purpose of what they would be doing. It enabled the teacher to plan this initial introduction very carefully, so as to develop her confidence in the use of tools in the classroom for the first time.

Classroom organization

In order to deal with the problems of starting in the classroom for the first time, the teacher first spoke to the whole class about safety in general and safety with the use of the tools in particular. They were shown as a class how to use the basic tools. She then took aside two pairs of pupils and demonstrated the construction technique for making the boxes. The pupils were asked to record the stages of making the box as soon as they had completed any particular stage, since they would be responsible for demonstrating the method to another group. In this way each group of two was responsible for showing another group. This enabled the teacher to circulate, ensuring that each subsequent group had understood the instructions, and to observe whether they were using the tools correctly (especially the try-square used for marking at right angles to an edge. (See the Appendix). Any questions would, initially, have to be directed to the pupils who had shown them the technique (AT 3).

The whole class was shown the variety of wheels available, and they were asked to consider at least two ways of attaching axles and wheels to their structures. They had to test their solution and demonstrate their findings to the rest of the class (AT 1, AT 4).

Throughout, the pupils listed the materials used, discussed these together as a class and made a chart on which they described the material, what tools they had used to work it, how it had been cut or glued, and any specific words they thought appropriate (e.g. 'bendable', 'tearable'). This chart was added to throughout the year.

The next major task was to decide how to decorate the body. The teacher showed the class one or two possible ways of doing this, but the class decided to use scales made from shiny paper, making their own templates and sticking the scales on to the covered structures individually. It became clear that this was a time-consuming task, and many suggestions were offered to resolve this, including 'paying someone else to do it'. The suggestion that was finally adopted was that the scales would be mass produced by the class. The operation was subdivided into smaller tasks (i.e. drawing around the templates on different coloured sheets, cutting the drawn shapes, gluing on to large sheets of card). The class split up into different tables to perform their part of the whole operation. All segments were soon covered with multicoloured shiny scales. The result was an articulated, segmented and very colourful dragon body, with fins attached along the spine for added effect (AT 2, AT 3).

All the class were introduced to some basic electrical circuits. One group constructed the head using papier mâché, and another used their knowledge of circuits to put bulbs in the eyes. The experimentation with syringes provided the pupils with an idea for using them in the head. The top jaw moved up and down (AT 2, AT 3).

Alongside this work, the teacher was developing a series of activities in anticipation of the planned dragon book. These included a number of paper engineering problems where pupils worked in card and with levers to make moving pictures. Pop-up and paper engineering books were brought in, read and experimented with. (Some starting points are suggested in notes for teachers in the Appendix).

Other language work was carried out around the theme of dragons. Dragons in mythology were studied, and the varied significances of dragons in different cultures were compared. Dragon stories were read and rewritten. Issues of stereotyping were raised and discussed and the pupils were confronted with a series of different dilemmas:

- Why should it always be princesses who are saved from dragons by princes?
- What if the dragon that had to be slain was the last dragon in the world? Do we not have a duty to protect it?
- Are all dragons seen as evil? How different is the Chinese dragon from the Welsh one? One parent argued that their child should not be involved in a project on dragons, since their personal religious beliefs stated that the dragon was a symbol of evil. This

moral/ethical dilemma was discussed in the class and the social issues addressed.

The pupils worked on their own stories and entered these into the computer, so that they could edit them. The final versions went into a class book on dragon stories alongside their moving pictures.

The description of the project highlights how the teacher was able to achieve the particular classroom activities she set herself initially, and how, at the same time, she was able to organize the introduction of the tools in a systematic manner that proved educationally useful by encouraging listening and speaking activities for a particular purpose.

The project was also used to raise a number of moral and social dilemmas, through the choice of stories read, issues raised and consequent rewriting. The ways in which dragons were depicted raised different cultural interpretations of the dragon.

The materials chart that the teacher initiated with this project and continued throughout the year enabled the pupils to add new materials, methods of working them, gluing them, describing and classifying them. The Appendix includes a possible framework for organizing materials within the class and the school. The work on materials in the project links with science work. In terms of science process skills, pupils had to carry out detailed recording, solve technical problems (attach wheels, joint structures, paper engineering), experience problems of friction and spacing. They had to discuss these problems in groups of twos, fours and as a whole class. They had to test their solutions and report their findings (AT 4).

Specific areas of knowledge included: electricity; properties of air; pneumatics; force; energy; mechanical movement; and materials. In Chapter 5 the relationship between science and technology is discussed in greater detail.

Topic: Chinese New Year

School: Alexandra Junior School
Teacher: Ms Jacqueline Buxton
(Key Stage 2: Year 3)

In a school quite separate from South Haringey Junior School, and not knowing that another dragon was being put together in another Year 3 class, this teacher had decided to initiate the project of making a dragon with her class on the occasion of the Chinese New Year. The Chinese New Year was to be part of work on China in general. Whilst Ms Chapman had responsibility for science, Ms Buxton was the Art and Craft Co-ordinator. Her additional objectives for this project were restricted by the limited time available to do the work, and were mainly concerned with highlighting the different graphical recording and art activities that could be developed from the work.

The construction stages planned for this project can be summarized as follows:

1. Show the pupils how to make a simple rectangular frame. The class were divided into friendship groups of two, and they were encouraged to decide their own sizes, given a maximum length and width.
2. Ask them to devise a method for attaching wheels and axles.
3. Place these in order of size and work out, as a class, how they would link these. The largest would be kept for the head, and the smallest for the tail.
4. Agree, as a class, a method to make up the body, and then build this on the frames.
5. Each pupil would draw their ideas of how either the head or the tail should look. These drawings would be discussed, and a design selected in the light of what emerged from the work.
6. All supporting material would be placed in a large class book (1.50 x 0.75m).
7. Decide on a front cover for the book that would involve every pupil contributing to it. (The teacher had planned to introduce the class to potato prints and had hoped to use this technique for the front cover in some way.)

Classroom organization

Once again, the biggest concern of the teacher was introducing the class to the use of the tools. They were very enthusiastic and keen to start, and the teacher was the first to initiate the work in the school, though she had not had any class experience of working with tools beforehand. In this case, however, the teacher was able to call on the support of an additional adult. She decided to borrow some more tools and introduce the whole class at once, allowing them to handle the tools and make their frames as a whole class. Since constructing the frame was a simple exercise, this proved successful in satisfying their immense enthusiasm and excitement (AT 3).

The pupils quickly completed their frames. Many groups worked so efficiently that they were able to complete two frames of differing sizes within a morning session. Having started the work with the whole class, the teacher was able to set supporting work (recording, drawing, measuring) so that there were smaller groups working on the next stage (AT 2).

After a demonstration of the use of the hand-drill, a description of how it works, and how to change the drill bits, the pupils were set the problem of working out how to attach the wheels and axles to their frames. A variety of materials were made available for this purpose, including possible connectors, such as clothes-pegs, biro holders, dried-up felt-tip pens with the core removed, screw eyes, and metal staples (AT 2).

Alongside this technical problem solving, pupils were asked to make careful and precise observational drawings of the tools they had used. The first drawings they produced had taken little time, and it was clear that they had not put much effort into their drawings. This was discussed with the class and the shape of each tool was pointed out to them. They were asked to consider how these appeared to look from different viewpoints. Questions of perspective drawing, of how things appeared different to the human eye from what they actually were, and methods of estimation using a pencil were all raised to help the pupils draw what they actually saw. Finally, a demonstration of how a try-square appeared, as seen from two different positions, was drawn on the board to show the concepts involved. (see Plate 6)

The class were then asked to redraw the tools with great care, using HB or even softer pencils (keeping records of first, second and subsequent drafts, so that they could reflect on the progress achieved). Similarly, they were also asked to draw the basic structures they had built. The main long-term aim of these activities is to teach pupils methods of drawing three-dimensional objects, so that they can use these skills to help them model ideas graphically in the future (AT 2).

The class discussed, as a whole, how they would link the frames together to make the body of the dragon (AT 1). Next, came the

Plate 6 Drawings of tools

most important decision: how to create a body for the dragon that would sit on the constructed wheeled frames? The class had constructed bases of different sizes which they had placed in order of size, from the largest at the head to the smallest for the tail. After a lot of discussion and many offered solutions, the class decided they would use welding wire to form 'wagons' for the body. This was a development from a previous project they had done. It allowed them to construct a rounded, articulated form by bending some metal rods (metal clothes-hangers or welding rod) to construct a skeletal series of loops, which could then be covered (AT 2).

Each group decided what length of metal rod they would need in order to make a suitable loop. They then transferred these measurements and cut the rod. After drilling holes in the correct position to locate them the rods were bent and fitted into the holes.

The next stage was to cover the hooped body structure. The pupils used strips of green tissue paper dipped into a Gloy paste (wallpaper paste would also be suitable). Any excess was removed, and the strips were placed over the hoops. By using coloured tissue paper, the work was complete after drying without applying additional colouring. The number of layers used gave an overall translucent effect (AT 3).

The whole class were then asked to think of designs for the head and tail of the dragon. This exercise generated a great deal of discussion. It was also very important in terms of drawing technique, as the children had to represent imagined three-dimensional images of the head in two-dimensional form (see Plate 7). This creative drawing contrasted with the previous experience of carefully recorded and structured observational drawing (AT 2).

On completion of the project, the written work, drawings and story writing were mounted together in a very large class book aptly titled 'The Dragon Book'. The front cover involved the pupils in making a series of potato prints, using different colours and basic shapes, as a means of patterning a dragon picture enlarged from a drawing. The results were impressive, because of the ingenuity of the stories and the quality of the final drafts of the recorded drawings, and the lively images of dragons' heads (AT 4).

The outcomes of these recording exercises are useful in the following ways:

1 It is often difficult for pupils to plan out solutions to problems and be able to communicate these to others before constructing and building what is in their minds. The teacher used the graphic recording exercises as a means of building up the pupils' confidence in drawing techniques, to enable them to start to lay the foundations for a vocabulary of graphical representation methods for future projects.
2 Careful and detailed observation work is still required in many areas of the curriculum and especially in art, science and mathematics. These drawings also place children in a position where they can consider concepts such as proportion, estimation and angles. These concepts can be highlighted, depending on the age of the pupils and the situation in hand.
3 The above drawings have the function of recording the work – in terms of the tools used and the different stages of working. This kind of work runs alongside the detailed instructional work (written and cartoon work) that the pupils have done.
4 Pupils are able to relate more easily to observational drawings of objects they have constructed themselves or tools they have used.
5 The contrast between the imaginative expressive drawings of the

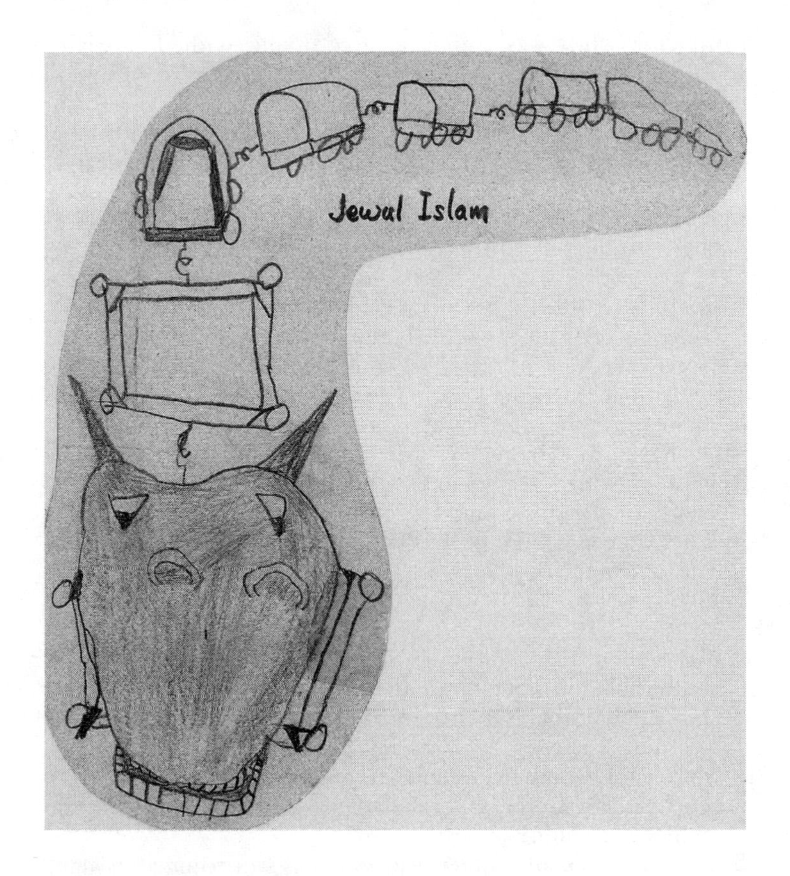

Plate 7 Dragons

heads and tails and the precise observational drawings highlights different aspects of art education and allows pupils to engage in both. The teacher also planned for a specific printing technique to be used within the overall project.

The end products of the two classes were very different. The teachers had approached their respective projects in different ways. However, both teachers agreed on one common factor. Both projects resulted in a joint class product that could not be split.

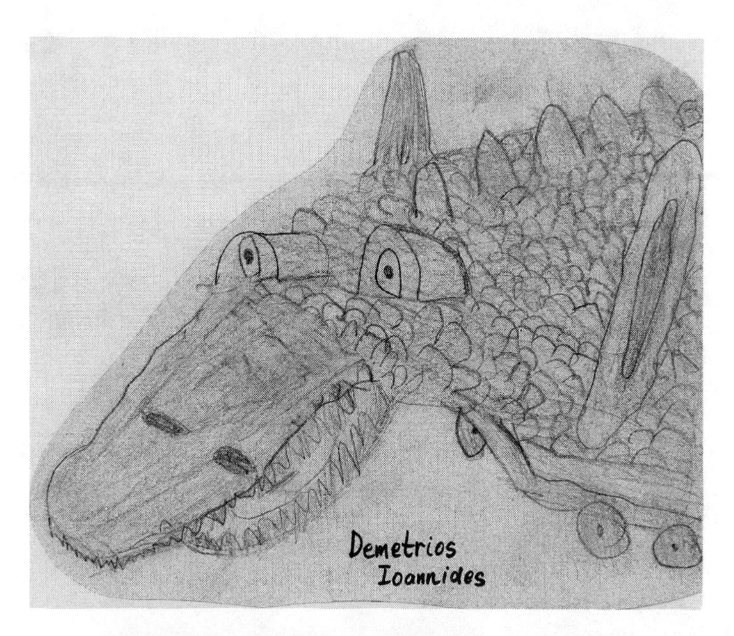

Both classes had developed a close affinity with their respective dragons, and there was a strong sense of joint ownership of the work. This caused great problems for the teachers, since they were unable to take any decisions on parts of the work without the respective classes discussing it and agreeing it. This was time-consuming, since carpet discussions were needed and votes had to be taken on all 'dragon decisions'. Teachers were not allowed to add anything without consulting the class.

However, when this problem was discussed, other teachers pointed out that the very process of social decision making was perhaps the most important outcome of both projects. Added to this, the pride and sense of common ownership that had developed was the mark of success of both dragon projects.

All three projects in this chapter were done with Year 3 pupils. However, other teachers have shown that it is just as possible to tackle similar construction techniques with pupils in Years 1 and 2, and with excellent results.

These methods of organization from a similar starting point are three of many. Other strategies could include involving parents, team teaching, and using outside visitors and support staff. The secret of the success of the three examples was the depth of planning by the teachers concerned, so that they could cope with what they felt was the challenge of starting with a very keen, enthusiastic and often over-excited class, whilst being personally concerned about how to organize and control the whole initiative, and achieve worthwhile educational results.

4

Links with mathematics

In this chapter, two design and technology projects are described that highlight specific mathematical concepts and knowledge. The first is part of a wider topic on masks, and involves pupils in designing and constructing nets to make masks. The second is building a scaled version of the school staffroom.

All the case-studies described in this book involve pupils in measuring, estimating, hypothesizing, and handling mathematical data in practical situations. The practical problems offer pupils realistic and meaningful contexts in which to handle a variety of mathematical concepts. Pupils quickly realize the importance of measuring, precision and accuracy in order to ensure successful results.

The main objective in this chapter is to show how it is possible to develop design and technology projects that serve to highlight particular areas of mathematics (e.g. nets, use of scales for enlargement and reduction, concepts of length, area and volume), whilst continuing to address attainment targets and programmes of study in design and technology, science and language.

Topic: 'Masks'

School: South Haringey Junior School
Teacher: Ms Maria Theodosiou
(Key Stage 2: Year 4)

The activities described in this project are part of a broader scheme of work which is shown in chapter 8. The class were asked to design three-dimensional masks using simple three-dimensional shapes. The masks would sit over their heads. Each mask would incorporate a mechanical movement or an electrical circuit using bulbs and/or motors. The pupils were encouraged to look at robots for inspiration.

The teacher's objectives were as follows:

- Develop the pupils' knowledge of two- and three-dimensional shapes.
- Develop their understanding and use of nets.
- Enable the pupils to decide a shape and design the net necessary for it. The teacher was experimenting to see how far this would be possible.
- Enable pupils to classify shapes (2D and 3D) according to commonly agreed criteria.
- Encourage the use of appropriate technical terminology.
- Ask the pupils to record their processes, and list the materials they used.
- To assess their understanding of 2D shapes, and their knowledge of the properties of these shapes.
- To explore the applications of nets in the pupils' environments.

The class had had significant previous experience in construction with timber, as well as other materials. They had handled the tools and were confident in using them to solve practical problems.

The teacher also wanted the pupils to develop the ability to plan and predict some of their projects before actually making them, so that they would start to work towards achieving specific goals. They had to be able to plan out their masks in terms of shapes as well as size before starting to make them.

A series of activities were devised to introduce the specific mathematical concepts:

1 The pupils were given two nets to construct that had been drawn up in 2-D.
2 They had also been engaged in pattern making, using basic shapes (square, triangle, hexagon), using a computer program.

Before commencing the main work with the pupils, the teacher reviewed the pupils' knowledge and grasp of the properties of two-dimensional shapes (e.g. triangles, quadrilaterals, squares, rectangles, heptagon, hexagon, octagon). Greek-speaking pupils kept reminding others of the correct terms. The teacher used the feedback and discussion sessions as a means of developing a

Plate 8 Making masks

common understanding of the criteria that would be used to identify and classify shapes (see Plate 8).

The pupils were asked to describe a particular shape until no other shape would fit the description. A game was played to encourage them to think precisely and therefore formulate the characteristics. Every time they gave part of a description, the teacher would show them a shape that fitted their description, but was not necessarily the shape envisaged. In the following example the shape is a square; the pupil's description is followed by the teacher's response.

- has four sides/edges – draw a quadrilateral
- has four corners – draw a rhombus
- the corners are 'straight' (after questioning, it became clear that they meant right-angled) – draw a rectangle
- the sides are equal – draw a square

Extra work was done on diagonals and the number of internal angles. There was also discussion on whether a shape was regular or irregular.

The class used the three-dimensional solids from mathematical kits to identify and describe the properties of the solids. They were also asked to construct a net for these solids. The teacher had anticipated this would be a lengthy process, but was amazed to find some pupils had worked out the nets of pyramids by turning the shapes on paper in less than five minutes. They were asked to produce two samples so that the work could be mounted flat and as a constructed net. The intention was to build a whole series of these, so that the class could hypothesize on the rules for constructing a net of one's own choice.

The next activity involved the pupils in identifying the simple solid shapes in a variety of objects brought into the classroom. Some of these were basic three-dimensional shapes (e.g. drinks cans); some (compound shapes) were made from a variety of shapes (e.g. winc bottlc). Different packaging examples were brought in (pyramids and prisms, papier mâché boxes in the shape of a heart, and similar containers were made available).

There are many ways of classifying three-dimensional shapes, and these were discussed. Pupils were introduced to the technical terms that arose from their descriptions. A discussion followed as to what constituted a prism, and what constituted a pyramid. Through a process of discussion and elimination, pupils were able to highlight a number of characteristics, as in the following examples.

A Prism:

- is a three-dimensional object
- it had a bottom which we called a base
- it had edges coming up from the base corners
- these edges were parallel to each other
- if the above edges came up 'straight'/perpendicularly/at right angles to the base, the prism was called a 'right prism'

- the prism was named according to the shape of the base; i.e. if the base was a triangle, it was called a 'triangular prism'; if it was a square, it was called a 'square prism'; if it was a quadrilateral, it would be called a 'quadrilateral prism'; if it were a hexagon, it would be called a 'hexagonal prism'
- if it seemed that a slice was 'lopped off' it was called a 'truncated prism'.

Similarly, pyramids were defined as:

- a three dimensional object
- it had a bottom which we called a 'base'
- it had edges coming up from the base corners
- these edges went up towards a point which we called an 'apex'
- if the apex was directly over the middle of the base, it was called a 'right pyramid'
- similarly, a pyramid was named according to the shape of its base
- similarly, if it was 'chopped off', it was called a 'truncated pyramid'

The second activity, pattern making using the computer, resulted in repetitive patterns around a central axis. This was a useful exercise for developing the notion that a circle was a many-sided shape. Through this activity pupils had been introduced to the notion of infinity, and the idea that a circle was a shape with an infinite number of edges. Thus, a cylinder was a special case of a prism where the base was a circle. Similarly, the pupils were able to identify the cone as a special case of a pyramid. They used the formulas they had devised to define a cylinder: as a circular prism, as well as an infinite sided prism.

Each group of two pupils was given a variety of objects, and asked to see what three-dimensional shapes they could identify and name, explaining why. The build-up work proved very valuable in this instance. It was also interesting to note how they had used the 'rules of classification' to build up the names of the objects they had been given. Those who finished classifying their objects were given others, more complex, or involving compound shapes.

When the class came together to classify their shapes, everyone had something to offer. Squeezy bottles were classified as 'right cylinders': a box of Maltesers was classified as a 'right regular hexagonal prism': a box of chocolates was described as a 'right truncated rectangular pyramid'. A discussion ensued when one group took a constructed net that had no labels and called it a 'rectangular prism', whilst another group called it a 'square prism'. The differences arose depending on how pupils had actually held the object, and according to what they decided constituted the base. One pupil asked if a fifty-pence piece was a pentagonal prism or simply a pentagonal shape. She finally decided it was a pentagonal prism with a very small height. The highlight was reached when the pupils decided that the papier mâché box in the shape of a heart was called a 'right heartical prism'. Often the pupils would forget one of the terms ('truncated' being the most frequent, and the teacher would motion to indicate an elephant's trunk as a reminder).

At this stage pupils were asked to consider what use they could find for constructing nets. The first response was sweet boxes. Packaging was discussed to some extent, and the various shapes they had seen were recalled. Pupils found it difficult to think about other ideas. They were asked to consider the packaging they wore to come to school. Attempts can be made to classify these shapes (a sleeve is a truncated cylinder). The manufacture of tins is also done from sheet metals. The use of 'real life' examples proved useful in stimulating them to look around for further examples.

The nets they had constructed from their mathematical blocks were looked at and reviewed. The aim was to show pupils the similarities between the nets of prisms on the whole and how these differed from nets of pyramids. The class would be able to devise certain rules for constructing nets.

The exercises above had been seen as an introduction to designing their own nets and to their choice of sizes. The play with technical terminology and rules of classification had fascinated the pupils to such an extent that work developed in this area. Visualization games were played, where pupils had to see the image in their head and answer questions like 'how many sides does a hexagonal prism with top and base have?' 'how many edges do you get in a square pyramid?' They had done so much work with three-dimensional shapes that it was possible for them to spend a short time visualizing the object and coming up with an answer, not based on guesswork.

The class was divided into groups of twos, and they discussed the task of designing and making their robot masks. They were given options of what materials they might wish to use and were introduced to corrugated plastic sheet, available in different vibrant colours. (It is used to construct cases, and is also now used as estate agents' signboards. It can be cut with scissors, a Stanley knife or a paper trimmer and can be scored and bent.)

The pupils were encouraged to look at sheet materials, so that they could develop their nets and build these up by scoring, folding, bending and gluing. However, many pupils insisted on building a basic framework in wood and then covering each side with sheet material, and finally adding basic shapes as nets constructed in card for the eyes, nose or other parts (AT 3). Before starting to do the work, they had to decide certain measurements and basic shapes, record these decisions, and then plan how they would construct them (AT 2).

Pupils who insisted on building a timber frame in the shape of a square pyramid found they had a lot of problems joining the pieces at the apex. Filling the angles helped, but it was difficult to keep them in place while the glue dried. A glue gun may have been useful at this point, but the school did not have access to one. Finally, a putty was found that hardened when exposed to air. This was used to fill all the areas at the apex. It worked adequately. Two groups used corrugated plastic sheet and this proved successful (AT 3).

The practical work of using and joining different materials raised the issue of adhesives, and the question of which were most suitable in each case. Pupils were able to experiment and investigate these. They also completed a materials chart/table where every group included the materials they had used and the approximate quantities as well. They used these charts to reflect on the overall resources.

During this constructional work, the pupils designed their particular electrical circuit, incorporating lights for eyes, or rotating bow-ties or antennae on the heads, including a simple switch device (AT 2). Some groups decided they would opt for a form of mechanical movement. In one instance, they constructed both eyes as two connected pulleys. When an axle was turned behind the head, the two eyes rotated simultaneously.

The pupils were also asked to record some part of the process/project where they had encountered a problem and to try to reflect on this by writing a report on the problems they had and hov they handled these. These provided very interesting reading to the outsider (AT 4).

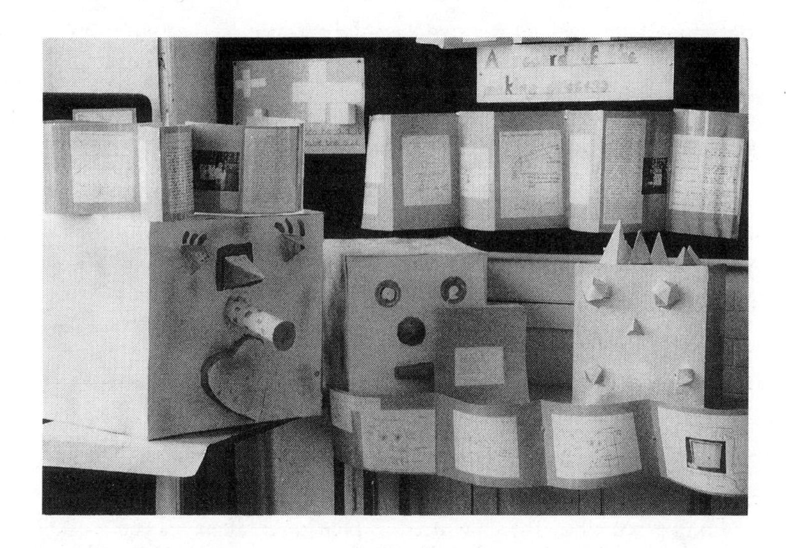

Plate 9 Robot masks

The absorption of this class with games was developed even further. Throughout the year, they had been engaged in making various masks (see Plate 9). They had also been building up a dictionary of materials and tools, and mathematical and other technical terms. The class devised some simple reading games, where the drawings of, for example, tools were also labelled. These were used with younger pupils when they used the tools for the first time.

The initial aims and objectives of the teacher were very ambitious; i.e. to ask the pupils to decide what shapes they wanted and then construct the nets for these. One of the objectives was to see how far pupils would be able to comprehend the concepts involved and then apply them to their own designs. Part of the proof of understanding would be whether they were able actually to construct what they had planned. The results were very successful, showing varying degrees of complexity and application.

There are many applications for constructions with nets; for example making puppets, packaging, making hats. These ideas can be used with Key Stage 1 as well as Key Stage 2 pupils, with changes to suit the levels of the pupils.

Throughout this project the class were aware of the purpose of the variety of activities; i.e. to work towards a knowledge and understanding of two- and three-dimensional shapes. It is useful to reflect on the overall activities, in terms of mathematics attainment. Table 1 on page 54 highlights the main ones tackled.

Topic: 'The environment'; measurements and scaling

School: Rokesly Junior School
(Key Stage 2: Year 5)

The topic was the environment, in a Year 5 class. They had previous experience with the use of tools and working in groups to solve technical constructional problems. The tools were part of the classroom equipment, and pupils had access to them as and when the need arose.

The aim of this project was to broaden the nature of the work they were involved in, and to develop more open-ended problem solving opportunities for the pupils. The class was divided into different groups tackling different projects around the school environment. This project arose out of the desire of the staff to consider changing the staffroom and reorganizing it. This was put to a group as a problem they could tackle. A model of the staffroom was needed in order to try out different plans, and get a sense of the overall atmosphere (AT 1).

Three pupils went and drew up all elevations in the staffroom as well as a plan. They had to devise a variety of means to work out the height of many objects. They designed their own measuring instruments, such as one metre measure sliding inside another so that the device could be extended to reach the ceiling or other out-of-reach objects. These measurements were transferred on to large sheets of sugar paper.

The pupils then had to decide the sizes of the model, and the concept of scaling was introduced. They chose the scale they would work to by a process of elimination. Half-size would be too big: similarly, quarter-size would not be able to be carried through the door. So, finally 'half a quarter–an eighth' was decided. Whilst the teacher tried to direct them to a tenth scale, they did not want to do this since, having tried out two measurements, they thought the model might end up too small. They then had to divide all their

Table 1 Mathematics attainment targets addressed through the work

Activity	*Mathematics*	*Attainment*
Describing 2-D shapes Describing 3-D shapes Classification of 3-D shapes	Shape and space	Recognize and use properties of 2-D & 3-D shapes
Devising criteria for classifying shapes Devise criteria for constructing nets	Decision trees (Handling data)	Represent and interpret data
Angles, edges, measuring Rotating shapes/repeating shapes symmetry/pattern making	Shape and space	Recognize and use transformation
Constructing nets from solids	Applying mathematics Shape and space	Use shape and space in data handling in practical tasks
Designing and making own nets	Using and applying mathematics	Use number, algebra and measures in practical tasks
Recording nets Recording plans and results Visualization exercises	Using and applying mathematics interpreting information presented orally	Use number, algebra and measures in practical tasks
Materials chart/making charts/reading charts/accessing info.	Handling data	Represent and interpret data
Discussion about infinity	Number	Understand number and number notation

measurements by eight. They started to do this by long division, but soon realized that a calculator would be much faster. The measurements were placed next to the full-size measurements on their drawings but in another colour (in order to distinguish clearly). They were asked to check that these measurements seemed correct by approximating the answers. They were using fractions, estimating results, using a calculator and working in decimals (AT 2). They then proceeded to build the basic framework with windows and doors to scale. They reinforced the structure in places and covered the walls with card (AT 3).

The decision was then made for the wallpaper design for one wall. This they showed to some members of staff who refused to accept the design. A discussion followed on the role of a designer in leading the design process and providing a variety of options and finally devising a means of evaluating the response. So they proceeded to draw five different designs and find out the preferences of the staff. They were questioned and their preferences were placed in order. The pupils collated this information to find out which design had the greatest number of points (AT 2).

The project continued with finding methods to print these wallpaper designs to full scale, and looking at some printing methods (polystyrene shapes cut and mounted on card as stamps, simple screen-printing techniques). The pupils also built the furniture to the appropriate scale (AT 3).

A very important element of all this work is the time spent gathering around and describing the work done, the problems faced and possible solutions. This also enables other pupils to offer solutions that the group may not have considered. On one of these occasions, this group had built the framework of the staffroom and covered it in card, and they explained that this was a model of the staffroom 'an eighth its size'. One person said that it could not be an eigth of the size of the staffroom. The whole group were asked to consider this and decide individually if this was the case or not. A vote was taken. The pupil who had voiced doubt explained why 'if you put eight of those staffrooms in a line you would not get the size of the staffroom'. The pupils whose project it was were very quick to answer that you would need to place eight staffrooms along the width and upwards, i.e. that every measurement had to be repeated eight times over. The group worked out how many models they would need to fill the staffroom ($8 \times 8 \times 8$). The teacher then directed the pupils to consider how many model windows would be needed to make up the actual windows and the whole group considered these questions.

Through the process of discussing and showing others, very important concepts were being explored; i.e. scaling, area and volume. What was particularly interesting was that the class had actually done a lot of work on these concepts with boxes and cubes, using mathematical blocks from the maths corner earlier on in the year. The three pupils who participated in this project were able to understand these concepts in a very practical manner. The girls who made the staffroom were confident that they had scaled down the staffroom and were able to substantiate this to the rest of the class (AT 4).

It is clear that a project involving pupils in making scaled models of a room or building, for the purpose of redesigning or planning furniture, would also involve pupils in a great deal of practical mathematical application, dealing with number, measurements and handling data and information, as well as applying mathematics.

We have described two case-studies that have taken specific areas of mathematical knowledge (nets and scaling) to show how it is possible to devise work that is design and technology orientated, while addressing general mathematics attainment targets (such as handling data, using and applying mathematics and measuring) as well as the specifics of the particualr projects.

5

Science and technology

In this chapter we will focus on some projects that have been tried out in schools and which give rise to areas of study that are associated with 'science' and/or 'technology'. Whilst educationists throughout the country have continuously acknowledged the importance of science education at primary level, there have been problems associated with implementing good science teaching. One of the biggest problems has been to build the confidence of primary teachers so that they feel they can handle science methodology, concepts and areas of knowledge.

In recent years, a number of primary science schemes have been published, aimed at helping teachers to focus on the science in topic work, or to show methods of science testing and give ideas for science activities that could be tried in classrooms. These schemes have led to varying degrees of success, dependent on how they are used. However, science education is now firmly on the agenda of primary schools, with the understanding of the importance of the role of teachers in providing science activities in their classroom.

There have been massive developments in technological education in primary years, but the research and implementation is not as advanced as in the science area of the curriculum. Teachers have engaged in technological activity in their classrooms for a long time, but have not necessarily recognized it as 'technological', nor felt able to draw out the technological concepts involved – until recently. Educational Support Grants (ESG) for funding science and technology in primary schools have led to an explosion of work in this area. There are many examples of excellent technological activity, but we are still unclear about the potential for pupils' conceptual development and capability, given planning, progression and continuity from a very early age.

We also feel that there can be confusion about what is meant by technological activity and its relationship to science. In this chapter, we look at some of the relationships and differences between science and technology as part of a process of widening this debate.

This task is difficult because of the very wide nature of technology, and of technological applications and processes, let alone the differences of understanding about what is meant by technology. Technology can be identified very broadly as activities that provide solutions to human need (open-ended problem solving) or more narrowly as providing know-how in a wide range of areas (technical problems). However, it is important that teachers are aware of some of the arguments and discussions that occur in this field. This awareness enables the teacher to focus on different thinking processes that are placed within practical contexts that have meaning to the pupils.

One of the most common misconceptions about technology is that it is a form of 'applied science'. This is based on the idea that scientific theory explains causes of phenomena, and that technology is the process of finding applications for that theory. However, Black spells out the differences when he is quoted as saying: 'Applied Science poses the question: "Is there a useful application for this Science?" whereas, technology asks: "Is there a solution to this human problem?" ' (TES 1989). These may seem simple pedantic differences that have no relevance in classroom teaching, and especially at primary level. We think this is not the case. Educationists believe that children learn from the practical experiences they are engaged in. They are also aware of the importance of making the work relevant and meaningful to the pupil.

The above notions can make teachers aware of the activities and/or problems these pose for pupils. 'Science experiments' can be meaningless to pupils unless they are placed in a practical context that they can identify with, and which gives their activities a sense of purpose.

For example, take the topic of 'Water'. There are many experiments around the themes of 'floating and sinking' or 'studies of materials in water'. These investigations are designed to stimulate pupils' skills in questioning, hypothesizing, testing, recording and evaluating. However, the technologists may argue that: first, the very reason for studying sinking and floating has not been raised; and second, that it is important to engage the pupil in activities that will enable the student her/himself to raise such questions as, 'what is the best material to construct "it" from? what adhesive do I need to use?' Therefore it is important to engage the pupil in a problem-solving situation that the pupil can identify with and recognize as problematic, as well as having a desire to resolve it. This engagement and stimulation will lead the pupils to solve the technological problem through a series of experiments, such as devising fair tests.

Another danger of science experiments, or investigation simply taken from work-cards, is that they often do not take into account children's own sense of 'know-how'. It was surprising to see the amount of 'knowledge' displayed by a group of top infants whose current topic happened to be 'Water'. They were set the task of making boats they would use in their school pool. When questioned about the shape, they were adamant that the boat had to be pointed at the front in order to 'make it easier to cut through the water'. They substantiated this by bringing in further examples that they had thought about from their swimming lessons to prove the point. Is it necessary for the pupils of this class to engage in a 'fair test' experiment or investigation that would prove which boat-hull shape in balsa wood is the most efficient?

The pupils arrive at school with an existing language expertise. The role of the teacher is to work on widening pupils' experiences in order to build on their capability. This is similarly true with technology and technological problem solving. Pupils had existing expertise of building, changing and constructing, using the materials around them. It is the role of the teacher to recognize, identify and build upon that technological know-how.

What we hope the above will raise is a series of questions that relate to the importance of starting points, stimulating the pupils' interests and putting them in situations where they can start to provide answers to their own questions and highlight the role of technological activities within the framework of the primary curriculum. Sparkes discusses the similarities and differences between science and technology, and analyses some of the processes inherent to both:

> Both fields (science and technology) use mathematics in order to make progress in their respective fields. Both are creative – though in different senses. Both theorise, though technological theories (information theory, control theory, network theory, etc) are theories about effective 'doing', rather than theories

> about causes and reality. And, of course, technology makes use of scientific data and theories just as science makes use of technology to arrive at its data and theories.
>
> (Sparkes 1987: 6)

However, he also raises some distinct differences between science and technology that are of interest, especially when they relate to different processes inherent in both fields:

	Key activities in Science	**Corresponding activities in Technology**
a	explanation and prediction	successful products
b	discovery	invention
c	theorising about causes	theorising about processes
d	analysis	design
	reductionalism (i.e. making distinctions isolating phenomena by controlling experiments and removing unwanted environment if necessary)	holism (i.e. bringing many different analyses bearing on design problems dealing with complexity and only controlling the influences
e	search for causes	search for solutions
f	study and research for its own sake	study and research for pursuit of only as much accuracy as is necessary for success
g	reaching correct conclusions based on accurate data	reaching good decisions based on incomplete data

(Sparkes 1987:7)

It is useful for teachers to reflect on some of these differences and question whether they may lead to different teaching practices. Putting work in a meaningful context and working with technological problems that achieve some form of human purpose are starting points that excite and motivate pupils. These problems can then lead the pupils into areas of knowledge and methodological scientific processes that challenge them. The result of such an approach is that it is no longer necessary to set fairly arbitrary 'science investigations' and ask the pupils to hypothesize, predict, test, record and evaluate investigations they may not relate to or see the point of. This is an important concern for teachers when the tasks are laid out for assessment at the ages of 7 and 11. It is much more challenging to ask the pupils to devise a fair test within the project they are involved in, and then look at the degree of 'fairness', the planning and the method involved in carrying out the task.

A lot of work has been carried out on the development of good science practice in schools. The study of the practical implications of the theoretical differences discussed is still relatively new. However, it is a rapidly expanding area of the primary curriculum. These discussions are part of the process of debate on the relationships between science and technology. The next three projects have been chosen to show some different topics that have been used to draw out particular areas of technology as a focus for the work.

Every one of the projects illustrated involves pupils in ScAT 1, relating to the exploration of science, which accounts for 50 per cent testing at Key Stages 1 and 2. They also involve pupils in a study and manipulation of materials, where pupils can be investigating their properties. Many of the projects involve pupils in studies of different forms of energy, electricity and magnetism. The precise subject-matter of the topic can raise other science attainment targets, such as those dealing with sound and music (if making musical instruments), earth and atmosphere (if involved in a project on designing and making weather measurement instruments), and so on.

Topic: 'Mechanical toys'

School: Rokesly Junior School
(Key Stage 2: Year 6)

'Toys and toy making' can be a very wide topic that could be used to raise many issues for pupils. New toys and old toys; toys and the gender implications involved; toys from all over the world; surveys of toy preferences and asking why some are preferred to others;

what is the function of toys? different types of toys – water toys, moving toys, push along, pull along, toys that make sounds, electrical toys, mechanical toys – the list is endless.

Many different toy projects have been carried out by teachers involving widely differing design-and-make activities for pupils of different ages. We have used this example from Rokesly Junior School to illustrate an attempt to map progression in a particular class that had done considerable work with tools in the year previously, and where there was a specific attempt to look at the group dynamics and interactions arising from different situations.

With many of the projects we have used in this book, group dynamics have been a consideration for teachers at the project planning stages. With this particular project the class had fewer girls than boys and many of the pupils had worked in friendship groups in Year 5. This had resulted in separate gender groups where girls and boys had tended to work separately in groups of two. The plan with this project was to create mixed groups and to study the interaction of boys and girls and, if necessary, to develop assertiveness amongst those who needed it. Special attention would be focused on who had access to the kits and materials, and how the groups discussed and shared ideas.

The group of pupils had considerable experience in handling tools and an involvement in open-ended problem solving in Year 5. The main objectives in setting this project were to:

1 Look at progression (in terms of construction methods, applications of previous experiences, recording methods and similar skills).
2 Look at group interactions and dynamics with special interest in gender observations.
3 Develop in the pupils' planning and research skills prior to making things in order to anticipate possible problems. The purpose of this was to encourage them to locate possible design and constructional problems before embarking on the work, and to use their developed observational drawing skills to predict shapes and forms.
4 To develop pupils' knowledge and understanding of mechanical movement in relation to pulleys, gears, ratios, direction of movement, cams and followers, and control.
5 To establish log-books and diaries to record progress of the work, as well as recording particular experiences.

The pupils were asked to 'brainstorm' around the topic of toys and were given a short time to write down all their ideas with no criticism allowed. Discussion followed on some of the issues raised in relation to toys (AT 1).

Visits to the Cabaret Mechanical Theatre in Covent Garden were arranged. The pupils were divided into two groups, each group making two visits – the first simply to enjoy the experience of the mechanical toys, the second to choose two mechanisms and work out and record how they worked and how they were constructed (AT 4).

Back at school, they were asked to construct some of the mechanical movements, both those they had seen and also any others they wished to construct. They were given the use of all the school's Technical Lego. These pupils were experienced in the use of these kits and capable of manipulating them to solve their own problems creatively (AT 3).

Pupils were divided into different sized groups of twos, threes and fours (using the large Technical Lego sets). The class were told that the teacher would not be available to answer questions of how to solve particular technical problems, but would be asking groups about how they were developing their ideas, and how they were working together to solve the problems. In the meantime, she was observing the group interactions and dealing with the issues that arose from the groups. It is not our intention to attempt to provide specific examples of these interactions to prove particular points, but it was possible to make some observations of the results of the work.

Where there were equal numbers of girls and boys, the boys tended to take over the construction – sometimes not even listening to the suggestions being offered by the girls. The girls were encouraged to resolve the situation by insisting on having hands-on access as well. A better balance was maintained in the group of two girls and one boy. Sometimes the pupils were asked to change chairs and roles in handling the construction kits. The issues arising from the group interactions were discussed with the pupils and they were asked to record how they felt about sharing the tasks.

The three-dimensional results were creative and technically well-thought-out. One group made a bird with wings that flapped as a handle was turned (see Plate 10b) using the technical Lego to simulate a beak, wings and a tail. Another group made an alligator's mouth with teeth and eyes which opened and closed slowly, again made

Plate 10 (a) Panda

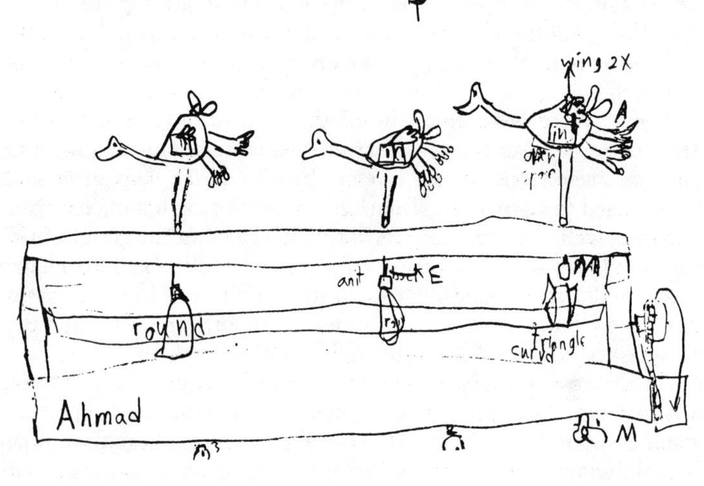

(b) Birds

(c) Animal with young

from Lego. These were drawn out carefully as a record of the work.

The next stage was to decide what the pupils wished to construct in their smaller groups of two. The general theme was animals and each group discussed their ideas and plans for making. Some groups were asked to make some of the simple mechanical movements they had seen at Cabaret, so that certain constructional problems could be discussed jointly and anticipated by all the groups. They were able to complete these in a day and these models were useful in the planning stages of their main project. In the following weeks the pupils built the basic structures to house their mechanical movements and built the animals and the surrounding environment using papier mâché and fabric (AT 2, AT 3).

The class had to keep a log of the work they were doing and the development of the project. This included diagrams, and at certain stages they were required to list the problems encountered and suggested solutions. As the school had access to a computer, the pupils were introduced to word processing and worked in their groups with the computer to produce their reports. Each group of two showed the next two how to use the word processor (AT 5).

Although the project took a long time to complete, the importance of the experience cannot be overestimated: it involved planning, problem solving, the enrichment of speaking and listening skills, as well as the skills of graphical communication and the quality of class 'report backs' at the end of the day was particularly high. The pupils had gained the opportunity to clarify their thoughts, to develop their confidence to speak in a group, and to develop their ideas as problems were shared and solutions presented. By the end of the project the class were confident in the use of terms such as 'cams', 'followers', 'pulleys', 'gears', 'gearing down', 'gearing up', 'cranks' and 'axles'. Two of the groups worked on adding electrical circuits operated by the movement of the handles. Their previous experience of working with circuits and switches had given them the understanding of these concepts and the confidence to transfer this knowledge to the current project.

The teacher also felt that it was a learning experience for her in terms of the practical problems of making some of the mechanical movements involved. It is important not to underestimate the time needed to plan, construct, redesign and reconstruct, let alone the frustrations when work does not go according to plans. However, it was also important to complete the work and stand back and view it in order to have pride in it.

Alongside this work with timber, card, papier mâché and fabrics, the class has been fascinated by some of the card models seen at Cabaret. Several of them had bought models and made them up. With other models bought by the teacher, they did some work on basic nets and developed this work to make articulated puppets from card.

Problem solving in this project was largely of a practical nature – with manipulation of kits and materials to achieve the idea, in this case to produce a mechanical toy. The project itself had enough input of ideas and suggestions, simply by using the visit to Cabaret as a starting point to determine, to some extent, the result. It is possible to approach a topic on toys in many different ways and we hope this will provide the reader with an insight for projects or topics that could look at 'wheels', 'mechanical movement' or other areas of science and technology from different viewpoints.

Topic: 'The transport problem'

School: Rokesly Junior School
(Key Stage 2: Year 5)

The following situation was put to a group of third year juniors engaged in a topic on transport.

A group of people living on an oasis have discovered that their supply of water is diminishing. As an interim solution, they decide to send a Morse code message to friends living on an island with its own fresh-water well. They ask them to send a tank of water as quickly as possible (AT 1). The friends need to bring the water up and fill a tank (the scaled down version is a Fairy Liquid bottle). This tank needs to be transported through a variety of terrains. The class is divided into groups to devise a means of lifting the water and transporting the tank across the listed geographical terrains. The groups were divided as follows:

1 lifting the water out of the well
2 crossing the sea
3 going across beaches and general terrain
4 through a river
5 through rocky areas
6 up a hill
7 across a ravine
8 and finally through the desert to their friends.

(Variations can be: through windy grasslands, across mountain ranges, down tunnels, across ice and snow).

This was the second project by the class involving the use of tools and general resistant materials. The first project was an introduction to the tools, by building simple chassis (similar to the base of the dragons in chapter 3,) working out how to attach wheels and then completing the construction by building transporters of various kinds (buses, racers, cars, land yachts, caravans and trains). The pupils had worked individually in the earlier projects, but were divided into groups for this project.

The main objectives that were planned for this project were as follows:

1 To develop some more open-ended problem solving whereby the

class would be involved with a variety of sub-projects along the theme of transport. The projects would come together for a particular purpose.
2 To investigate the skills and concepts, as well as attitudes, that would be encouraged through open-ended problem solving.
3 The different projects would raise a wide number of different investigations and areas of science and technology. How can the teaching situation be organized to make use of the different investigations? How can the teaching situation be managed to draw the common threads (e.g. use of different materials to suit different purposes, different forms of energy, the process of scientific fair testing)?
4 To study the different methods pupils employ for designing and solving practical three-dimensional problems.
5 To monitor group work.
6 The pupils in each team or group would be required to produce a group report.
7 The diversity of projects and the varying nature of the solutions would be used to encourage pupils to develop their research skills.
8 To look at methods of reporting back on progress.
9 To map progression from previous work, in terms of design methods, planning procedures and practical construction skills.

The first group, the well group, built a model of a well and proceeded to design a mechanical method for bringing up buckets of water to the surface. They had done some previous work with electric motors and wanted to motorize their designs. They were advised to try this out in Lego before proceeding to construct and change their models. They soon faced the problems associated with using a string on a fast running motor, when everything got tangled. They then had to spend some time coming to terms with some of the concepts involved in slowing down motors and gearing. They then had to transfer these concepts, learnt through the use of constructional kits, into a model made with found materials that would actually work (AT 2, AT 3).

The sea crossing involved experimenting with materials that would float and sink. Resource books gave them some good ideas, but they had to translate these into their own situation. None of the ideas in the books included carrying a container full of water. They also had to discuss how their container would cross the sea.

They finally decided to use sail power. Designing the sails, methods of fixing them, and steering mechanisms were some of the other challenges the group had to tackle.

Other groups faced general and specific challenges, such as: how sturdy did the machine that crossed over rocky beaches have to be? how can we protect the tank of water from possible damage along its voyage? what type of wheels would be needed for each terrain? what forms of energy would be suitable to power the vehicles?

The river group initially shared ideas with the sea group, but eventually decided that the problems were different. Rivers were narrower than the wide open sea. Different materials were used that raised their own problems. The adhesive they first used was inappropriate in water. Their model fell apart. Sails were inappropriate, and a decision was made to use paddles. Finally, elastic bands were used to power the paddles.

Another group had to consider the types of wheels they would need to help their transporter climb hills and mountains. How do we stop slippage backwards as the transporter goes up? What form of energy is needed to get the transporter up? How do we ensure the tank does not fall off? After a great deal of trial and error, they decided to motorize the vehicle as well as add a hook that could be thrown and hooked to something at the top of the incline. This would be part of a mechanical winch attached onto the transporter.

Moon climbing buggies and science fiction films were the initial inspiration for the first ideas of the group tackling the ravine crossing. They soon realized that it was very difficult to transfer these ideas into solutions they could make, given the materials and resources available. They then decided to resort to more simplistic solutions, whereby they would use gravity and angled ropes to send the tank from one side to the other. When trialling out these ideas, they found they were giving themselves further problems and had to devise solutions to the inevitable crash as the tank of water arrived on the other side. First-aid carriers were then discussed. The work of the water-well group and some research into other sources enabled this group to adapt their later solution and design a motorized cable car device.

It was fortunate that the pupils in the last group had the actual experience of someone whose car had got stuck in the desert sand, and who had had a lot of trouble bringing it back onto the road surface. The wheels became a dominant feature of the design of this vehicle. Protecting the water from the extreme heat was another one of the challenges they faced. Wide wheels, made from drink cans which were also sprayed proved the most successful in the end.

It will be clear to the reader that all involved enjoyed the breadth and challenge of this particular project. There was a great sense of fun throughout the hard work. It is also clear that the organization of the work, the preparation of resources, as well as coping with the imaginative ideas of pupils and helping them to translate these into practical success, is a major feat. Different groups covered different areas of knowledge, but shared common processes. However, the range of solutions and breadth of knowledge (water testing, sail designs, motorizing, wheels, friction, pulleys, steering) enabled them to be aware of how wide the possibilities are.

Observation of the different design methods used by the pupils proved, to the teachers concerned, that there is no one way of designing, and that they, as teachers, should be looking at the methods of organization they need in order to facilitate these different methods.

Many pupils designed by constructing models. These models were 'thinking models'. They lacked any rigidity, precision, durability, or, in many cases, any aesthetic consideration. Some may argue that standards of construction and the final product are not important considerations in such a situation, but that the thinking processes are more important. The product is important to the pupils. Feelings of achievement and success are paramount in building their self confidence. If they are able to improve on a model, then they need to develop their work further from a 'thinking model' to a final design. The teacher is the professional who is best able to decide the limits for her/his particular pupils. It is also possible to develop concepts of precision and accuracy by showing pupils the need for these in the context of the particular project and the demands for quality of production. Therefore, it is necessary for these models to be rigid, precisely made and well constructed, with special consideration given to the types of finishes suitable to each design. In this project progression in constructional techniques and standards was one of the objectives of the work.

The 'thinking models' were discussed with the different groups. Different testing methods were used, including discussion about the likely points of weakness: dropping the work a distance; putting weights on or pressure at certain points. Where necessary, groups had to rebuild their models. The management of these sessions proved very important. They were used to show pupils how they may proceed from the 'thinking model' stage to the next production stage, using skills that had been built up before drawing their models, planning the reconstruction procedure and sharing the tasks.

These briefing sessions were used to develop the group or team identity. Records of these were kept to be used in the final team report. Tasks were divided so that the report would be shared. A structure was provided for the possible layout of the report. A simple one is offered below:

- Group members.
- What was the overall problem?
- What was your sub-problem?
- How did you solve it? Explain with as many drawings as necessary.
- What materials did you use? Draw these.
- What problems did you have along the way, with planning and construction?
- How did you test your ideas and final solution?
- How did you work together? How did you share the tasks?
- What did you enjoy most about the work?
- What did you enjoy least about the work?
- What do you think of the finished model?

After the pupils had completed the work and produced all the documentation, the headteacher interviewed some of the groups. These interviews were taped and the recordings were used in an exhibition alongside the work to explain to parents and other visitors what they had done (AT 4).

Since this project, other teachers have used similar ideas for work, based on the theme of 'travel' and alongside work on *The Hobbit* whereby two young children are engaged in an imaginative trip across many lands. Their encounters and descriptions of the lands they travelled through provide the basis for a mass of creative writing, drawing and storytelling.

Topic: 'Water'

School: Coleraine Park JMI
Teacher: Ms Jenny Gaden
(Key Stage 1: Year 2)

The class had had some experience with the use of the tools. They had used timber and PVA wood glue. The teacher had been involved in an INSET course, where she was involved in open-ended problem solving, using the tools that would be available in the classroom. She had decided on the topic of 'water', and had anticipated it lasting a long time, so that the pupils could be involved in some in-depth study.

It was decided to set the pupils the problem of building boats. The class was divided into groups of twos. They were allowed to choose their partners, so that there could be a development of confidence. The pupils brought in lots of materials that are found generally: tops of boxes, meat trays, boxes of different sizes and other 'scrap materials', pieces of timber, some pieces of plastic, Fairy Liquid bottles. The teacher was also involved in finding appropriate waste materials. They had many books available with pictures of different forms of water transport, as well as other books on water in general.

They were asked to design and make some simple boats. They were also warned that these boats would then be tested in the sink to see how they floated and moved. The pupils would also carry on till they had produced boats that could float, and would be tested in a small swimming pool that the school had easy access to (see Plate 11). There would be a race to see which boat went the longest distance or the fastest across the pool.

The results they produced used different bases as the bottoms of the boats, and they built these up to simulate different parts of the boats. It is interesting to reflect on some of these bases. Some were polystyrene meat trays found in supermarkets; others were fairly solid tops of cardboard boxes – some where the glue was used to join and construct the top, others where the corners had been stapled together. Some of the pupils added pieces of timber to simulate different parts of boats.

Their boats were then tested in the sink and other large containers. Initially, they generally floated. However, they were left in the water for the weekend to test properly. On Monday

Plate 11 Testing boats in water

morning, the pupils were eager to verify their results. All of the boats, with the exception of some of the polystyrene based boats, had sunk.

At this stage, the teacher spent considerable time questioning the pupils as to the possible causes of the different failures. This was important in developing their ability to hypothesize as to cause and effect, and to build a resource of concepts and knowledge that would be useful for their future problem solving. Group discussion was important, and all the possible reasons were debated. The pupils then had to draw a picture of their model and explain how it had been tested, and what they had found out (AT 1). The results and explanations included:

- the glue used to join the corners together had come apart and the boat sank
- the cardboard tops that had been stapled kept their shapes but 'the cardboard got soggy and then sank'
- the polystyrene trays that had got some water in them had sunk
- polystyrene trays that did not get water in them stayed afloat

After a lot of discussion, the pupils developed concepts such as:

- we need to build the boats from materials that do not sink, and they must be tested for a long time
- the glue we use must be glue that does not come apart in the water
- the shape of the sides must be such that water does not easily get in

When questioned about the shapes of their own 'waste material' boats as compared with the other boats, they seemed very clear that a boat needed to be 'pointed' at the front, 'so that it could cut through the water and go forward'. They had made their boats with what was available, and that was generally rectangular (AT 4).

They were able to list, at this stage, what materials would sink and which ones would float. However, some of them wanted to test some other materials, particularly wood, since they hoped to build their next model in timber, using the 'wood glue'. They felt sure the 'wood glue' would be 'stronger' and stay together in water. For this purpose, they went about testing their hypotheses. They tried out different pieces of timber in water. They glued some pieces together, and then put them in water. A few other materials were also tested, and corrugated plastic was introduced at this stage.

It is important at this stage that the teacher is aware of notions of mass and density and some of the criteria that affect floating and sinking, so that she or he is prepared to direct children's questioning so that certain concepts are developed. Some of the pieces of timber they had used were fairly heavy. Some sank quickly. Others sank overnight.

At the end of their particular experiment the pupils were able to ascertain a number of points:

- heavy wood sank quickly
- wood also got soggy, then heavy, and sank
- the wood glue they used came apart in the water as well
- certain materials did not sink generally, unless water got into them and made them sink

The next stage of the project was to get pupils to think about their boats again, and help them construct shapes that might make their boat easier to move in water, as well as to make them out of materials that would float, and to use adhesives or constructional methods that would avoid problems of leakage and joining. The teacher wished to get the class working with the corrugated plastic sheets. The teacher also introduced thin plastic sheet as an experiment to see how well it worked.

The pupils were then shown a glue gun. It was pointed out that the instructions with the glue had said that it was waterproof. While they were shown the gun and how it worked, it was made clear to them that they would not be able to do the actual gluing themselves, because of safety considerations. The teacher would do this for them while they watched. The pupils were also told that they would have to construct their boat shapes in card first and then use the corrugated plastic to make the boat afterwards. They were shown two different approaches to constructing their boat in card.

- to design the base of the boat, and then construct a rim around it
- to draw the sides of the boats (two identical sides), and then place a base sheet between the sides

The pupils made their shapes in card and joined them together with adhesive tape. This work involved a lot of practical measuring and fitting together and it was an excellent introduction to and practical application of nets (AT 2). They were also shown methods of cutting, scoring and bending the plastic sheet. However, transferring the card measurements on to the plastic required accuracy. This notion was developed practically, as they saw the effects of messy cutting and the large gaping holes that would be left for water to flow in.

While the class started making the models in card together as a class, the development on to corrugated plastic sheet was done with smaller groups, so that time could be given for consideration of all the factors and to ensure care was taken in completion of the boat hulls. In the meantime, every group recorded the stages they had completed, and anticipated the next (AT 3, AT 4).

When the boat hulls were complete and tested for simple waterworthiness, the pupils were asked to consider different types of sails they might attach to their boats. They were asked to consider: how they would make the sail and what its shape should be; how they would fix the sail on to their boat. They were shown a variety of possible sail materials, as well as some constructional methods they might find useful: e.g. bending wire (coat-hanger, welding rods, etc.); joining garden cane to create triangular shapes, using rubber bands or string as a means of lashing or binding.

At this stage the class was becoming very excited at the thought of testing their boats in the swimming pool, as well as having the great race. It was important that this event was done as quickly as possible, so that they could enjoy the results. The teacher spent some time discussing with the group the rules of the race. A lot of discussion ensued as to how they would time each boat, and they tried out a number of timers. In the end, they decided that counting together would be the best solution in the given circumstances.

Evaluation and testing sessions are of paramount importance for developing concepts, and testing the results of previous hypotheses, designs and redesigning processes. It is a crucial part of any scientific, technological and design process. However, in this instance, it is fair to say that the afternoon spent testing the boats in the swimming pool with the pupils was, in the main, a highly enjoyable affair. The pupils were very excited and it was all they could do to wait their turn to blow their boats from one end of the pool to the other, as well as count for others. While the counting proved to be very irregular and the whole process could not be considered in any form or shape to be exemplary of 'fair testing', the session proved very successful, in that each group proved to themselves that they had designed and made a boat that travelled the length of the pool, worked by being blown, and could be used as a toy to play with in the future (since none of them fell apart), and was a product they could be proud of (AT 4). The teacher decided she would further test the boats when she took the pupils to a pond close by at the end of the school year.

While the pool race signified the end of this particular part of their water project, the class were to carry on with their designing and making activities by looking at different methods of propelling simple boat shapes across water. Different types of paddles operated by using elastic bands and inflated balloons, were among some of the ideas they used to propel their boats.

The three topics illustrated in this chapter have been used with pupils of different ages in other schools, with results that reflect the maturity and previous experiences of the pupils. However, the common thread throughout these is to develop pupils' self-confidence to plan, decide, execute, record and assess the work they have done throughout the whole project and any sub-problems the teacher has decided to locate and concentrate on, in order to highlight particular skills or areas of knowledge – scientific, mathematical or linguistic.

6

Links with information technology

The information technology (IT) profile component in the Technology Statutory Orders clearly states that IT capability should be developed through activities which arise in different curricular areas, and in this sense, IT applications are truly cross-curricular in nature. The editing and printing facilities of word processing packages and printers are one application for the English Statutory Orders; databases and measurement sensors are applications listed in the Science Orders; spreadsheets and simple programming are listed in the Mathematics Orders.

The educational potential of the use of computers across all Key Stages is developing very fast and it is often difficult to keep up to date with these developments. The aim of this chapter is to introduce the reader to some of the general applications of information technology within the design and technology profile component, and to one area in particular, namely measurement and control. We choose this application because of the excitement pupils get from working in this area, its links with the practical aspects of designing and making, and how the conceptual and manipulative skills involved can be further developed and applied.

It is possible to connect a variety of hardware devices to the computer that can be used to measure time, temperature or humidity. A buggy, known as a 'turtle', is one control application often used for work in mathematics. This turtle will move according to simple programmed instructions devised by pupils. It is also possible to use other hardware, referred to as control boxes, together with appropriate software programmes, to connect lights, buzzers, electrical motors and cassette players, controlling them in a series of simply sequenced programmes.

The main aim of this chapter is to illustrate ways of approaching computer control projects within a topic approach. There are many different control packages available for the different types of computers. The Micro Electronics Support Unit (MESU) has produced resource packs showing the availability of packages and options, together with excellent resource packs full of ideas and applications for use in the classroom.

While the aspects of control and measurement are indicated in the Science Programmes of Study and can be applied to general logical sequencing and decision making in mathematics, we have found Year 2 and Year 3 pupils able to use and design simple programmes that control a series of outputs (such as lights, buzzers and motors).

The three case-studies used in this chapter involved Year 5 and Year 6 pupils. However, they can be adapted for use with pupils of earlier years. They have been chosen because they present the reader with a wider range of topics (environment, robotics, and myths and legends), as well as a gradual build-up in the use of the control packages, whereby the pupils start with simple electrical circuits, design their own circuits and switches, and then apply these, first to controlling outputs (lights, buzzers and motors), and then develop these to incorporate a number of inputs (switches) that control the output sequences.

Before describing some case-studies of projects followed with pupils in schools, it may be useful to consider and question the introduction of these activities into the primary curriculum.

1 As an extension and progression of the practical work the pupils are engaged in. Computer control packages are an excellent means to introduce electronics to pupils through the use of simple electronic devices in the form of switches (Light Dependent Resistors [LDRs], tilt switches, reed switches, and microswitches are some). It also helps to develop the notion of logic gates by introducing pupils to the idea of making options (If this action were taken, then do A. Otherwise do B).
2 To simulate the function of robotics and other control hardware, that are an important application in the 'real' world.
3 As an excellent method for developing sequenced, logical thinking, testing and evaluating. This offers pupils an enjoyable and user-friendly introduction to the concepts of computer programming. This use is part of the Mathematics Attainment Targets and the turtle is one example of such a use.
4 The language needed to use the control hardware is very simple, young children can use control packages with a limited vocabul-

ary. Logo software is also available and can be used as an extension of the mathematical work.
5 Pupils enjoy working with the control package and making up their own programmes, seeing instant results, and solving technical problems.

Topic: 'The environment'

School: Rokesly Junior School
(Key Stage 2: Year 5)

This Year 5 class was introduced to the control pack as part of a topic on the environment. They had practical experience of working with the tools and materials and had been divided into groups to look at traffic lights, Belisha beacons, pedestrian crossings and street lights. This project had developed from investigative work in electrical circuits (other work related to this topic and class is discussed in chapter 4).

The stages of development could be subdivided as follows:

1 An introduction to electrical circuits. Making their own circuits.
2 Trying out available switches and making their own.
3 Programming outputs (bulbs, buzzers and motors) in simple sequences.
4 Using switches as input devices to control the bulbs, buzzers and motors.

Each group of two had been given a kit consisting of: electrical wire, electrician's screwdrivers, bulbs, bulb-holders, buzzers, motors, as well as a series of switches. These included reed switches, tilt switches, LDRs, potentiometers, microswitches, PTM (push to make) switches, and PTB (push to break) switches. The following chart describes the activities the pupils were asked to try out. On the right of the chart are some of the concepts the teacher anticipated the pupils would develop throughout the topic.

Activity	Concept
1 light a bulb	what is a circuit; a power source; load; conductors?

2 switch on a buzzer	the buzzers work when the current goes in one direction and not the other: concept of polarity
3 make the motor turn	
4 reverse direction of motor	polarity needs to be reversed
5 try the various switches	what a switch is; the range of different switches; electrical resistance
6 make a break in your circuit and try different materials across the break	conductive; non-conductive; semi-conductive materials
7 make your own switches	application of 5 and 6: introduction of soldering techniques
8 make a reversing switch	technical problem solving
9 light two bulbs	circuits in series; circuits in parallel

(See Electricity Resource Sheet, The Appendix)

After an initial 'hands-on' session, the work was recorded on the chalk board. This involved drawing the results of each electrical experiment accurately and writing comments. The purpose of this activity was to make the pupils aware, quickly, of the tedious nature of such a detailed method of recording, thus illustrating and introducing the need for abbreviations and symbols in circuit diagrams. Pupils used symbols to record future experiments.

The purpose of the above activity was to immerse the pupils in electrical circuitry and to illustrate the type and variety of problems they were likely to encounter. A substantial time was allowed for each group to explain to the rest of the class their particular findings and designs for switches or any interesting observations they had made. Good electrical connections proved to be very important to the success of some of their experiments and soon the whole class had been shown and had used a soldering iron (see Plate 12).

Having immersed themselves in electrical circuitry, the pupils then approached their chosen design task. Many designs for streetlamps were thought out and built. Table-tennis balls and

Plate 12 Soldering connections

Smartie tubes proved to be exceptionally good solutions for constructing Belisha beacons. Traffic lights of a variety of sizes were constructed as well as pedestrian crossings. One group had been to the library and found a book on building traffic lights from which they insisted on copying the circuit. As the drawings in the book were not clear and circuit diagrams were not included, the resulting model did not work. The pupils were frustrated by this, but when asked to take their model and work without the book with the knowledge they had learnt earlier, they were able to make it work very quickly. This clearly illustrated that, although books were important as a general resource for ideas, the pupils were hindered when they tried to copy the solution without understanding how or why the model worked.

The first group to complete their model was the group making Belisha beacons. These were subsequently used by the teacher to introduce the whole class to the computer control package. As this was the first time the class had used a computer, the excitement was great. The output controls of the package were demonstrated, with the construction of simple programs as the group experimented with their Belisha beacons.

The language used in the control program was very similar to Logo. These were some of the processes:

1 Switching on and off a series of outputs.
2 (a) Building a sequence, (b) ending a sequence.
3 Repeating a sequence.
4 Making a sequence recursive.
5 Testing a sequence.
6 Editing a sequence.
7 Naming and saving a sequence.

It was obvious that the desire for 'hands-on' experience was great, and, since this was the first time the school had had the use of a computer, there was some apprehension as to how it would be shared and used by the class. It was decided to take two groups of pupils, show them how to handle the computer, how to load the program and repeat instructions for using the control software. They then had the opportunity to familiarize themselves, before showing the next group all these procedures. With careful management of the groups with regards to time, and occasional teacher intervention to boost the information, all the pupils had an opportunity for hands-on experience of simple programming.

In the mean time, the pupils working on the lights and beacons had completed their models and worked on simple procedures on the computer. The procedures were thoroughly discussed, with some pupils acting as lights following given instructions.

Next, the pupils needed to use input devices. A series of switches were plugged into the input control sockets, enabling pupils to make 'conditional statements'. An example would be:

- if input on 1 then switch off 1
- if input off 1 then switch on 1

A tilt switch is plugged into input socket no.1 and a bulb is plugged into output socket no.1; i.e. 'if the tilt input is on, then switch the light output off and if the tilt output is off, then switch on the light output'.

This part of computer control work involved pupils in more complex programming, with greater discussion on procedures and involvement in evaluating success or failure of those procedures. Programming mistakes are more likely to occur and more difficult for the pupils to find. Hence, the language used between the pupils involved in the work is much more interesting. Listening skills become paramount. Making use of group problem-sharing is interesting and fruitful. In this situation, a group of pupils from the class assemble around the computer and the fault is explained. Often the solution is obvious to someone else not involved in the project. While the teacher may manage such instances initially, as time develops the pupils get used to the idea of going over to ask someone else to help them.

It is impossible to overstate the importance of these methods of working. They simulate real planning situations. They also highlight the importance of working in different ways: the divergent modes of thinking are very important in approaching a problem; convergent methods are needed for refining. However, it is absolutely necessary to be able to look openly and divergently at the situation throughout the stages. It is important we assist pupils in finding means to do so. Sometimes an insistence on going out to play and then coming back to resume work is one method of doing this. At other times, looking at the problem and discussing it with a partner for a period of time, without acting on any decisions, is another.

The types of problems solved with the use of control hardware were technical in nature, and it may be possible to list some of the areas of conceptual development and techniques needed and covered at this stage:

1 Using one program within another.
2 Building conditional statements.
3 Solving the technical programming problems that arise in the experiment/finding faults.
4 More complex programming.

The students were able, after a lot of work, to complete the following:

1 Sequence the Belisha beacons.
2 Sequence a number of different sets of traffic lights.
3 Sequence the pedestrian crossing (this program was later built into one of the traffic light sequences).
4 Light the street lamps in the dark.

The next and final stage of the project was to use these devices as part of a street scene where other pupils' work could be fitted (a house with its own lights, vehicles powered with a small motor).

Topic: 'Myths and legends'

School: Rokesly Junior School
(Key Stage 2: Year 5)

The same class had developed an interest in circuitry as a result of the work they had done in the above project and some others not mentioned here. They had a good grasp of what a circuit was, what a break in a circuit was, fault finding, circuits in series and in parallel. They had developed circuits in their heads. They were very confident with the use of the soldering iron.

In their first topic of the year, they had done substantial work on myths and legends, and especially stories relating to the Greek gods. They were building a class nine-headed hydra, using modelling techniques similar to papier mâché. Once again the class was divided into groups. They were shown a construction method using chicken wire to build up a basic shape of a snake head. They proceeded to build their snake heads (nine in all) and cover these with a material called Modroc (quick-setting plaster-coated scrim, which needs wetting to start the setting process) (see Plate 13).

The pupils decided they would want to have lights that flashed on and off for the eyes. They were able to do this as long as they were aware that they would have one battery to operate both eyes. This meant they had to apply their practical knowledge of circuits in parallel.

However, as the work on the computer control developed, it was decided that it would be more interesting to sequence the eyes to light up at different stages. The idea was further developed so that the nine-headed hydra would not start flashing unless its area and base was interfered with. The heads were attached together and a murky base of a swamp constructed. A switch (which would be the main input control device) was made out of a crab, so that if the crab was touched the different eyes in the nine heads (forming the outputs) would start flashing in a specified programme of sequences before they settled again.

This project developed as the teacher decided to try to attach a

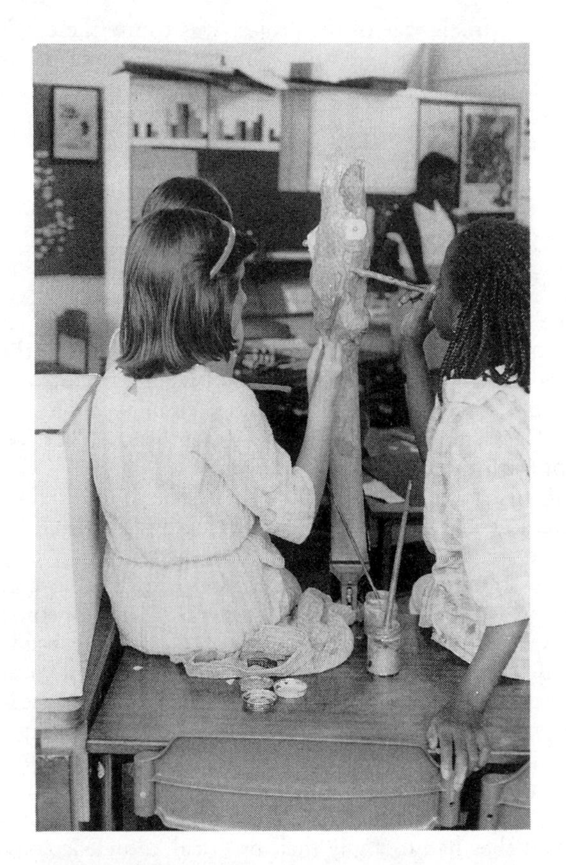

Plate 13 A nine-headed hydra

tape recorder to the control hardware, so that the hydra, when disturbed, would not only start flashing its eyes, but would also start a rumbling thunder series of sounds.

In order to achieve these sounds, the pupils collected a series of sound-making instruments (water, tins, rattles, blowing bubbles in water) and taped a sound sequence that simulated the sounds of a disturbed nine-headed hydra. The tape recorder was attached to the control hardware and became another output device that would also be activated when the crab was touched.

The whole programme was put together quickly, and it worked

relatively well. There were problems with changing the eyeball bulbs, fitting them in and checking circuits. What amazed the teachers was the persistence of the pupils in being able to locate these faults and find ways to resolve them.

The nine-headed hydra went to a series of exhibitions in the borough, and finally came back to the school. What proved particularly interesting was the future use it was put to. It was used as a way of introducing pupils to the computer control output package. Pupils would work in twos, decide on a sequence for the different heads to light up, and then work on achieving that sequence. It was easy to make up a sequence, since the heads were coloured in different ways, and were easily distinguishable.

Topic: 'Robots'

School: Rokesly Junior School
(Key Stage 2: Year 6)

At Rokesly Junior School there has been a lot of work done throughout the year groups with Technical Lego. There is a policy to develop it alongside the Basic Lego in the lower years. The pupils are always encouraged to show their work and explain how they did it, both at assembly time and often to another class. The Lego sets are well looked after, and each class has access to a number of sets. (See notes on use of Lego in the Appendix).

The MEP (Micro Electronics Project), now superseded by the MESU, gives, as examples, a variety of control project ideas. Using Technical Lego or cheap motors with tin cans and various other types of pulleys, pupils can build motorized buggies, which can be developed by attaching them to control packs and sequencing different movement patterns.

While pupils enjoy using kits to construct models, it is important that they are able to transfer some of their understanding into other materials, and simulate the same mechanisms in models of their own constructions. This transference of concepts into other media ensures a thorough comprehension of these concepts and their different applications. The Year 6 class at Rokesly had had considerable experience with the tools, and had been involved in a number of design-and-make projects. Their topic was 'robots'.

There were many other areas of the curriculum that the pupils were covering within such a topic, apart from the obvious technological implications. The most interesting were the number of robotic dances that they had devised. The joint work, co-ordination and partnerships involved could be the subject of a chapter on their own.

Alongside this work, the pupils were given a number of technical problems to solve using the Technical Lego kits. Lego I and Lego II kits were available, and one group had access to a pneumatic Lego kit. They were given a series of mechanical movements to try to build: an arm going around slowly; a robot moving backwards and forwards; pincer movements; movements simulating cranes; barrier movements, and many others. The teacher set these movements to pairs of pupils, and waited to see the response. The aim was to introduce the pupils to the use of the motor, the need to slow the motor down by use of gears and pulleys and simulate different motorized movements.

The pupils also brought in their own toys from home: especially space buggies with motors; transformers of all varieties; and other kits, either with good examples of mechanical movements or with some aspect that was motorized. This inspired their projects, so that the rotating arm became a camera carrier moving in all directions, and the vehicles became part of a space buggy.

Using their Lego models, they devised a series of programs for their robots to move in a variety of sequences using the control equipment. These were shown to the others, and the programs explained. The majority of pupils were more interested in transferring their kit ideas into projects that could be taken home, rather than going on to use the inputs to any degree of sophistication. However, the outputs were used fully, with flashing lights and buzzer sequences to accompany their various models.

The principle of gearing down was quickly established, and the pupils became fairly adept at understanding the functions of gears and cogs. Some used pulleys to slow their motor down. Positioning the motors was found to be an important factor in their designs.

Other activities during this period included keeping a log of the work in progress. This was completed every time the pupils spent a session on their robotics project. The log included: a review of what had been achieved; personal comments about the progress; sketches of the relevant details; and finally an idea of the next stage.

It is possible to use the control hardware and software for many different projects, such as sequencing motorized machines in the fairground or looking at different machinery that may need to be sequenced. A sequence from a play or nursery rhyme can be built. The teacher who is interested will find the resource packs that are available within their LEAs and Colleges of Education very useful and helpful. Some ideas, used with teachers, are shown in the Appendix.

7

Whole school development

The previous chapters have shown a variety of case-studies which have included different methods of starting out with the use of the tools. Many teachers had been on in-service training (INSET), aimed at developing their personal self-confidence in tackling open-ended practical problems, involving handling the tools and areas of technology they might not have experienced before.

One of the recurring comments of teachers on the courses is that they fear starting projects which tend to be open-ended in nature. They are concerned about their possible inability to answer practical questions that may be raised by pupils. This worry is compounded by the many organizational problems arising out of the use of tools in the classroom for the first time.

The aim of this chapter is to look at some of these concerns and address some of the issues involved. Once again, two case-studies are used to illustrate how school staff decided to work collaboratively to overcome some of the fears, and to work towards building a school policy based on the teachers' experiences of curriculum development in the area.

The first example shows how one school (Down Lane Junior School) embarked on curriculum development with a clear strategy that accounted for the need to develop staff confidence whilst addressing the educational needs of the pupils in the school. The second school example (South Haringey Junior School) shows how a school started to tackle the issues of developing a school policy where work had been going on for two years.

Many concerns are raised by teachers embarking on the use of the tools for the first time in their schools. It is possible to identify some of these, which tend to be organizational in nature:

1 Working with tools and materials in general.

2 Starting to use these with pupils in the classroom situation for the first time.
3 Coping and making use of the immense enthusiasm that is generated by starting this work with the pupils who have had no previous classroom experience.
4 Dealing with the issues of safety.
5 Choosing starting-points and challenges that are relevant to the classroom topic and are considered appropriate design and technology activities.
6 Planning a balance of time allocation for the design and technology activities, to ensure that other curricular activities are addressed.

Chapter 3 showed possible ways of organizing pupil groupings, resources and support to initiate practical work in the classroom, using a particular constructional technique to build frames and boxes that were incorporated into the class topics ('dragons' and 'farms'). Chapter 2 showed how a teacher introduced a reception class to the tools and materials, using collages as a theme. Each teacher will finally devise some particular organizational starting-point that is best suited to her/his circumstances and needs.

The case-study below aims to demonstrate the usefulness of working together as a school staff, and to show how a co-operative effort could reap far wider benefits for staff and pupils. It also demonstrates that, while design and technology activities can be time-consuming, they can also act as a focus for many other areas of the curriculum.

Design and technology as a focus for language work

School: Down Lane Junior School
Headteacher: Ms Irene Lomas

Down Lane Junior School has ten classes in its school. Its pupils come from a variety of cultural backgrounds and twenty-six community languages are spoken throughout the school. At the beginning of the year, seven of the eleven members of staff were relatively new to the school: four probationers, one teacher who had just completed a probationary year, a new deputy headteacher, and another experienced teacher new to the school.

The staff of the school had decided that they needed to concentrate on reviewing the language policy in the school. Through discussions with language advisory teachers, they decided that the best means of studying language development was through a policy of 'design and technology', since they believed that these activities could give rise to a variety of different types of language activities that could be seen to have a 'real' sense of purpose for the pupils. A study of these different language activities would be made following the work.

The plan of action involved two members of staff attending INSET courses. They would be responsible for initiating work in their respective classrooms, so that they could act as role models for others to observe and question. This period would also be accompanied by staff workshops for the whole staff, to introduce the tools and discuss some of the practical problems. Other curriculum meetings highlighted language issues arising from the work.

The second stage would be to involve every member of staff in the use of tools in her/his classroom. This stage is described in this chapter. This period would be assessed and would lead to a development of more open-ended problem solving as the confidence and expertise of the staff developed.

After the initial workshop sessions the staff decided to try out frame constructions in their classrooms. They felt this was the easiest and most structured way of starting the work, and a method they felt they could handle. The main aims of this school exercise were as follows:

1 To introduce pupils throughout the school to the use of the tools in a safe and organized manner.
2 To enable this introduction to be shared amongst all the pupils in the school.
3 To give teachers a common experience that would enable them to share ideas and discuss these relative to their particular classes. This would initiate collaboration and discussion.
4 To decide some common language activities that could be developed throughout the school.
5 This initial work would be seen as the first step towards developing more open-ended problem solving. While the box construction involved no designing, there would be some technical problem solving involved.

6 Any 'box' construction must be part of the class topic, i.e. there must be a purpose to these structures. The teachers, among themselves, saw the school topic as 'Boxes', but each class had a different topic. Figure 2 shows some of the different ideas that had been used or emerged.
7 The practical projects would be seen as the focus for different cross-curricular activities.
8 Certain activities would be common to all classes (e.g. observational drawings, language activities), so that a basis could be established to study progression. Figure 3 lists these common activities, and it is useful to reflect upon them.

Since speaking and listening skills were an important part of the school's language review, it was decided that pupils would be grouped in twos to encourage discussion through the practical work and the problem solving activities involved. This would also enable the staff to make a study of the group dynamics and interactions. Special thought would be given to the criteria for deciding the different groups, in terms of friendship groups, gender groups, mixed-ability groupings and language diversity. Every group of two would be involved in making a box of some form. This would be used as a method of introducing the tools to the classes.

Each class and group would be involved in making observational drawings of the tools used, the stages of making the boxes, and the finished product. The aim of these activities would be to encourage careful and detailed observational drawing. The choice of drawing medium could be varied (charcoal, pencil, ink, crayons, paints).

The detailed observational drawings of the tools would involve the pupils in a study of the tool, to see what it felt like; its weight, what it was made of, and how it worked. They were also asked to describe the tools and their function in their own words.

The drawing of the three-dimensional structures would develop pupils' three-dimensional graphical capabilities, so that they could use these learnt techniques as a means of communicating ideas graphically in the future.

A lot of preparatory work would be needed to discuss drawing techniques: apparent and true lengths of measurements as they appeared to the eye of the observer; drawing what is seen and not what is known (a lot of work has been done in this field – and the book *Drawing on the Right Side of the Brain* may be a very useful resource for teachers).

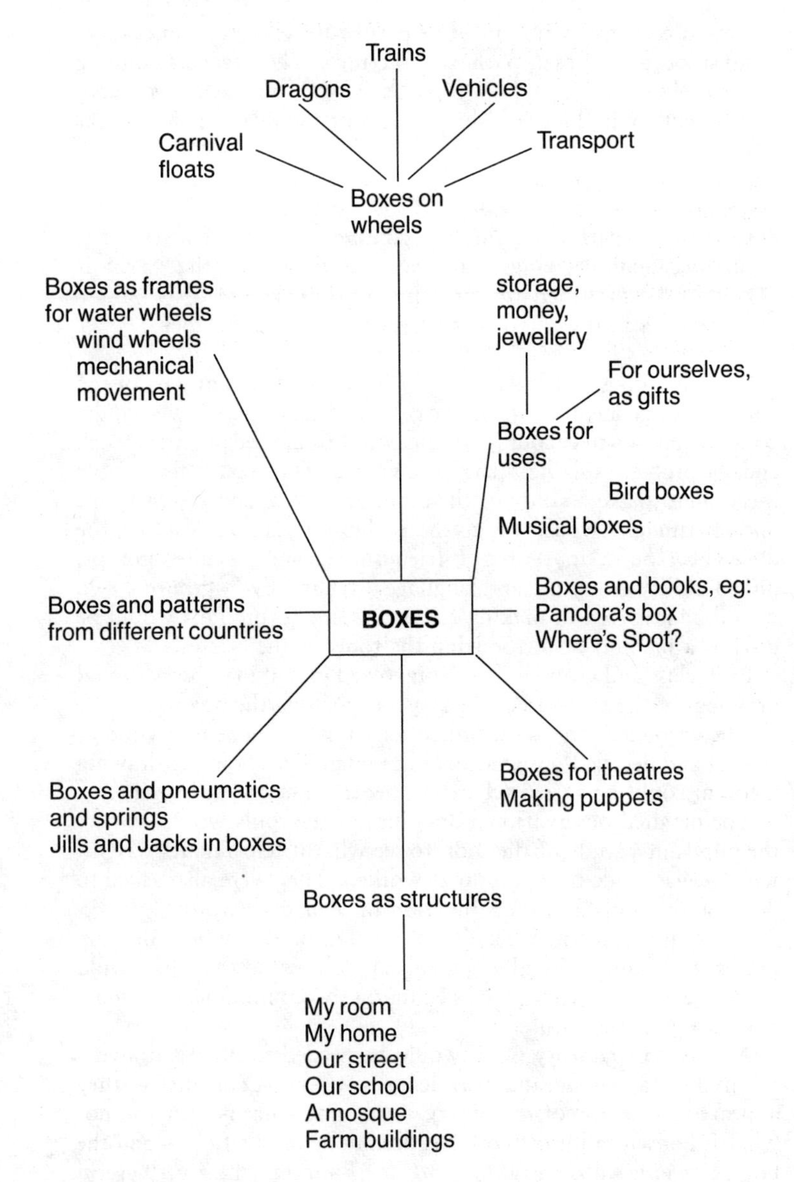

Figure 2 Possible applications of the Jinks box

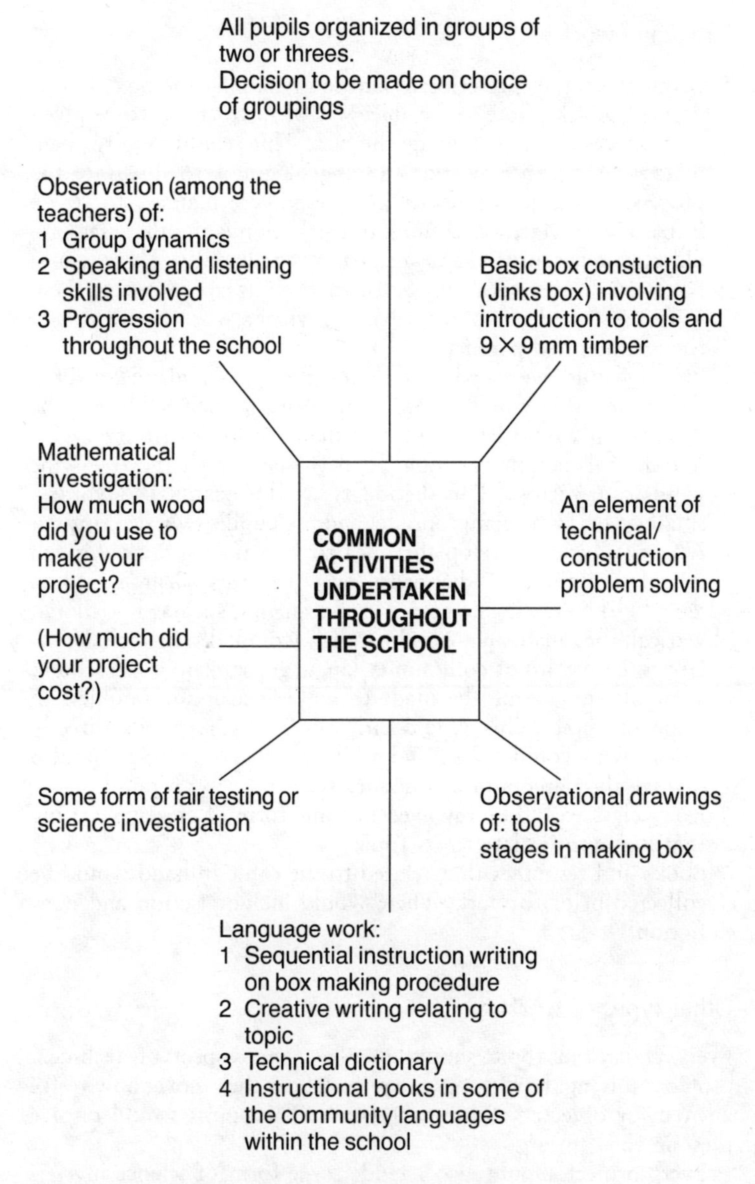

Figure 3 Common activities throughout the school

Language work

1 Given the rather prescriptive nature of making the box, it was decided to ask pupils to produce sets of instructions to be given to others on how they made the box. This would involve them in logical sequence writing, adding diagrams to illustrate the process. A variety of reader audiences were planned for these instructions: Martians coming to earth; pupils in other classes – older, younger, or the same age; parents and adults. These would be tested on the other classes for their accuracy. (Some of them were also tested at a parents evening when a workshop was held for one hundred parents.)
2 Pupils would be asked to describe the tools and write out a description that would include comments on their function. The description would be included in the technical dictionary.
3 A technical dictionary would be developed of all the terms the pupils came across. This dictionary would include: the word; a small sketch, where appropriate; and the pupil's own description. It is possible to develop this idea so that pupils work through school, building this dictionary over the years. Words used to describe the drawing of corner card triangles, such as horizontal, vertical, diagonal, square and parallel, would also be included.
4 Given the wealth of community languages spoken in the school, some attempt would be made to make instruction booklets in some of these. This would either involve the pupils directly (those who could speak and write another language) or also involve the parents in a collaborative project with pupils.
5 Every class would be involved in some form of creative writing that was appropriate to its topic.
6 Books and resources that related to the topic in hand would be collected prior to work. These would include fiction and non-fiction.

Other types of work

Every classroom topic should involve some aspect of technical problem solving: how things are joined; how they move; how to fix an item or object; finishing the work. The pupils would resolve these in their groups.

Every project should also include some form of science investigation or fair testing. The whole school would be involved in a practical mathematical investigation that answered the question: how much wood did we use in our project? Some classes would be able to respond to: how much did it cost? This exercise was important for the school to decide how much timber it would need for the following year. The aim of the exercise was to involve all the pupils in the school with a joint task that had a useful function. Compiling cutting lists and pricing design products are activities that have a 'real' function. This project would involve pupils in measuring and costing exercises with a purpose. It was also important that these exercises took account of progression in measuring.

In order to do this, it was proposed that Year 3 pupils would use some unit of measurement, and measure accordingly. There would be a great deal of discussion beforehand, leading up to the concept of a 'unit of measurement'. The Year 4 pupils would be encouraged to look at the ruler as a tool for measurement. The Year 5 pupils would also measure their timber accurately, but would have to make some effort to work out how many lengths of timber they used. This would include looking at the scrap pieces left over. The Year 6 pupils would conduct a similar exercise to Year 5, and be responsible for collating all the information in the school.

Teachers have tried different and equally successful organizational methods to conduct such an investigation. It is possible to ask pupils to discuss and decide on the method by which they will calculate the amounts and then to work on the problem independently, thus providing a means of cross-checking. On the other hand, it is also possible to have the group work together throughout the exercise, and have it cross-checked by the teacher or another group.

Individual classroom topic plans of activities

After a lot of deliberation, a number of topic plans emerged. The following pages describe some of the topics, together with a very brief description of the planned activities, as well as those activities that emerged through the development of the work itself. Additional activities that could be incorporated have also been included. The first topic includes the common activities which have been abbreviated in the following plans.

The aim of the plans together with the very brief descriptions is to highlight how design and technological activity in primary schools is not only justified and essential, in its own right, to a pupil's development, but can also be used as a focus for the topic work, in order to plan relevant and challenging activities in other areas of the curriculum.

'Ourselves'/'Our town'

The pupils had been engaged in considerable work on their families, themselves, and their homes. As a group they were asked to consider what type of a town they would live in (for the variety of classroom activities generated, see Figure 4).

After establishing the types of buildings this small town would need, each group of two pupils was asked to take on responsibility for the design and building of a model for one of them. They drew what they thought the structures would look like. Alongside this work, they constructed some three-dimensional forms using waste materials.

They were shown the constructional technique for building the frames. The measurements from their temporary models were transferred into the timber using a variety of measuring methods: comparative measurement; use of Unifix cubes; string and rulers. They soon found out why accuracy was important, and methods of achieving this were discussed.

They had to work out how they would cut the timber for their roofs and attach these to their existing structures. They had to work out how they might enable their doors to open and close. In this respect, they all had some specific technical problems to solve in their individual constructions. Those groups that had larger buildings soon found out that they required strengthening struts, and many small problems arose which the pupils had to resolve themselves.

This class spent a major part of their time finding materials to simulate building fabrics, and great care and detail was placed on the finishes to their various building structures, involving them in marbling techniques, colour mixing and brick patternwork, as well as investigating textures in a wide range of materials.

'Growth'

In this project, the class had been planting seeds and bulbs and

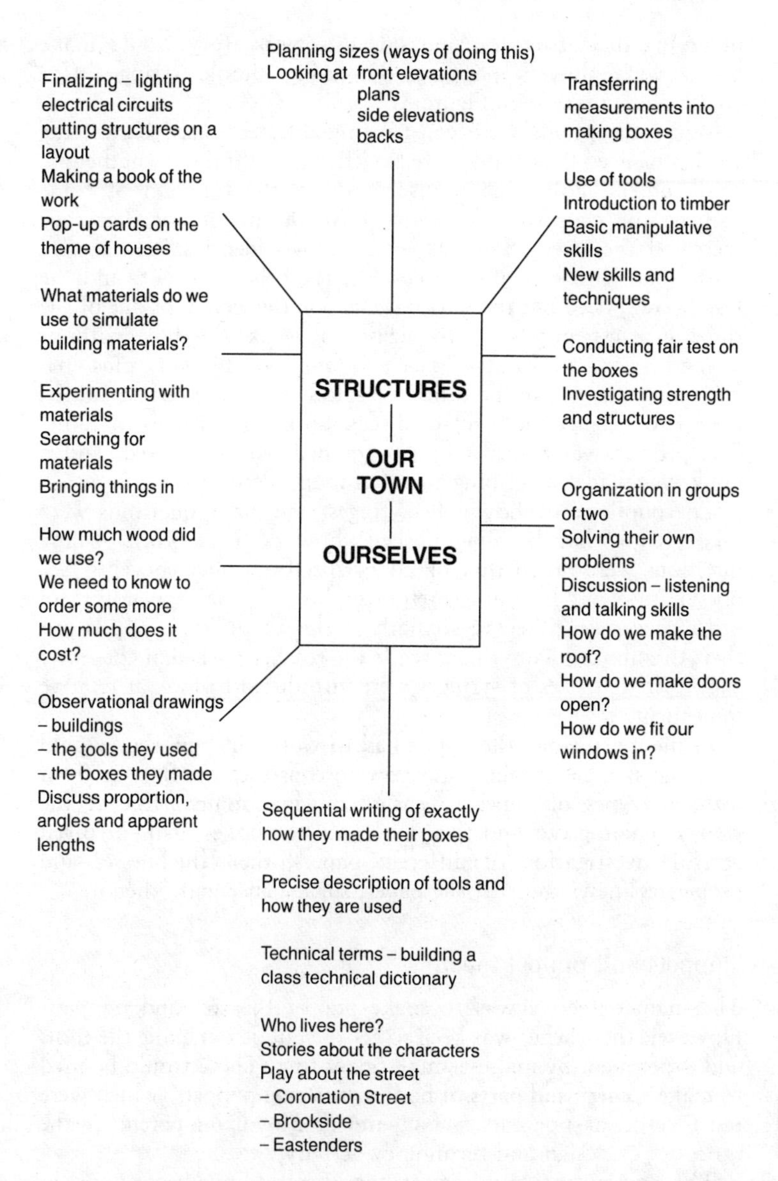

Figure 4 Breakdown of classroom activities planned for a topic 'Our town'

observing their growth. Alongside this work, they would make boxes with flowers popping out using simple syringes (for classroom activities see Figure 5).

While the pupils were constructing different sized boxes, the teacher noticed that many were forgetting to add the strengthening card corners. Following a discussion on the function of these corners, the class were asked to prove the hypothesis that they increased the strength of the boxes. Two identical boxes were made, one with card on the corners, the other without, and the tests were carried out the next day. It took the weight of one of the three heaviest pupils in the class to break the box without strengtheners, whilst the teacher (approx. 11 st.) plus ten encyclopaedias broke the other. Needless to say the testing caused great excitement, and a lot of discussion for many days to come. The process was recorded, photographed, weights noted, and a book was made, describing the afternoon's activities.

The pupils were shown the syringes, and many questions were asked; e.g. what is being pushed? how do they work? Many questions were raised that highlight specific science concepts (see pneumatics and hydraulics resource sheet in the Appendix). In order to demonstrate the strength of the air pressure, pupils put their thumbs on one syringe, while the teacher pushed at the other end. Different sizes of syringes were introduced to look at relative movement.

In the mean time, the pupils had to work out how they would hold the syringes in place and how to construct the flowers from different types of paper to ensure the maximum effect of the flowers coming out and then back into the boxes. This involved careful investigation of different papers, their thicknesses and properties (news-sheet, tissue paper, sugar paper and others).

'Puppets and puppet theatres'

The main objective was to make puppet theatres and puppets. However, the teacher was keen to get the pupils to handle the tools and experiment by making small boxes first. These would be used to make scenery and parts of bodies for string puppets (which were put together at a parents evening aimed at involving parents in the processes of design and technology activity).

The consequent building of the theatres involved pupils in planning sizes and considering certain functional details, such as

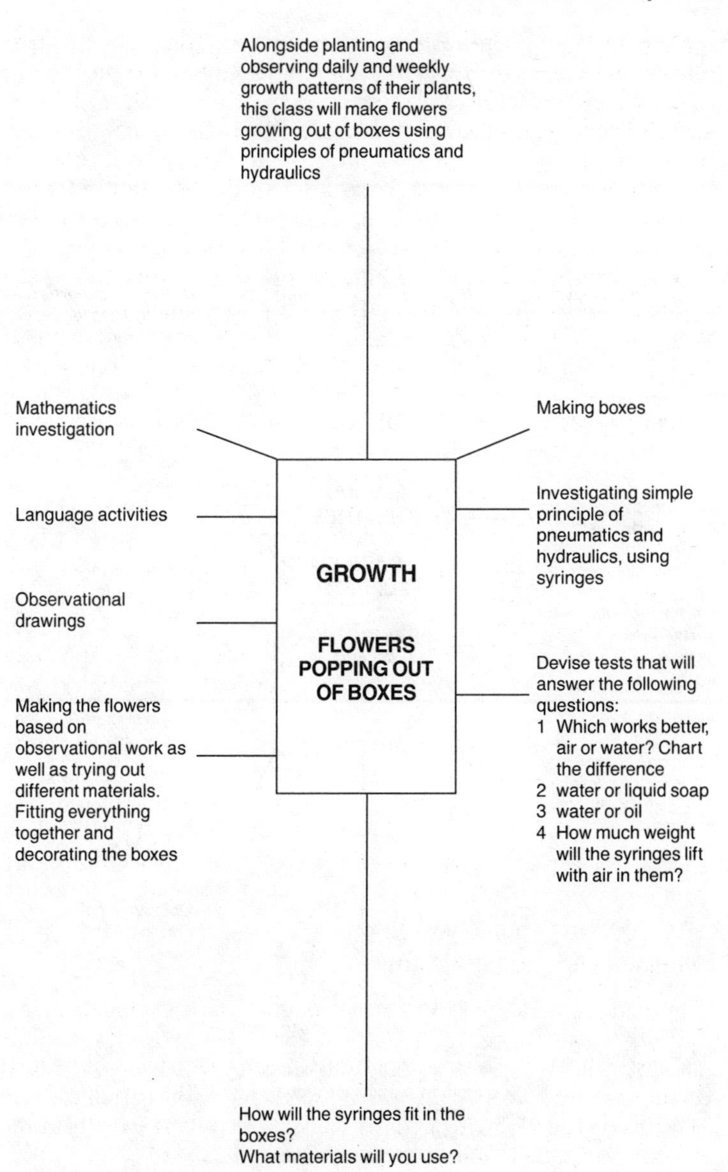

Figure 5 Breakdown of classroom activities planned for a topic on 'Growth'

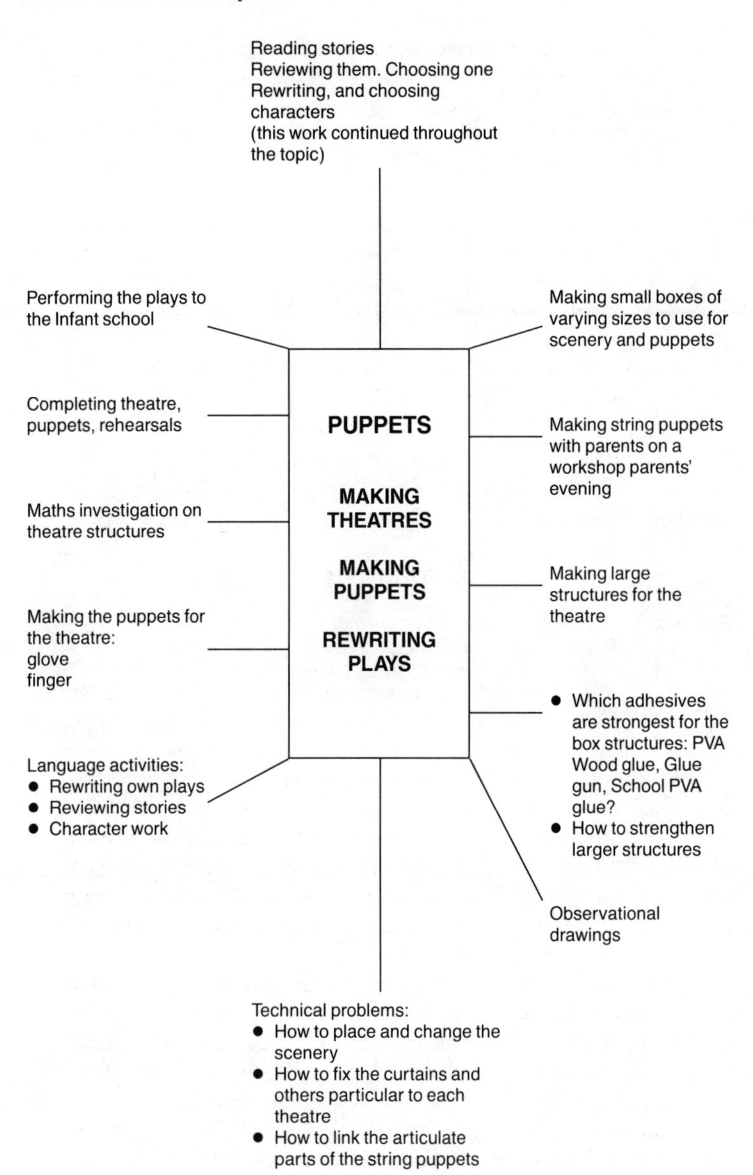

Figure 6 Breakdown of classroom activities planned for a topic on 'Theatres'

changes of scenery, stage design and opening and closing the front curtains. The theatres were to be used for glove and finger puppets in scales appropriate to their size.

The pupils noted that the teacher had brought a different PVA glue (Resin W PVA) from the one they generally used in school. They asked whether there was a difference. The investigation conducted by Year 5 pupils was reported to this Year 3 class, and they were able to conduct their own investigations relating to the different adhesives used for joining timber.

Such a project involved the class in story reading, reviewing and choosing their favourites, and finally writing these down in the form of plays. The work included visits to theatres and museums, as well as the practical work of deciding on characters and making glove or finger puppets. A number of art techniques were incorporated, as well as textile printing and fabric construction. Throughout the topic, the pupils were able to build a folder that described the materials they used. Finally, they performed their shows to an audience of infant pupils.

'Colour'

This Year 5 class had previous experience of the use of tools. They had also written precise sequencing instructions on how to make some paper engineering cards and applied these techniques to book making (see paper engineering resource sheet in the Appendix). It was important, therefore, to set them a more challenging situation that would further their capabilities. The topic was 'Colour'. Having read about and discussed the significance and traditions associated with the Hindu colour festival of 'Holi', the pupils considered what a colour festival in Haringey would celebrate and who it would involve. They were asked to design mechanical carnival floats for such an occasion.

The pupils visited Cabaret Mechanical Theatre in Covent Garden, London, for ideas. A number of visual examples were brought in from a local secondary school. They drew these and made observations and recommendations on improving them. They also used Technical Lego sets to simulate these movements.

They made simple box structures on wheels, that would contain mechanical movements to operate the figures and scenes on the tops of their floats. The technical problems and ideas were very similar to those discussed in chapter 5, and the pupils worked out different

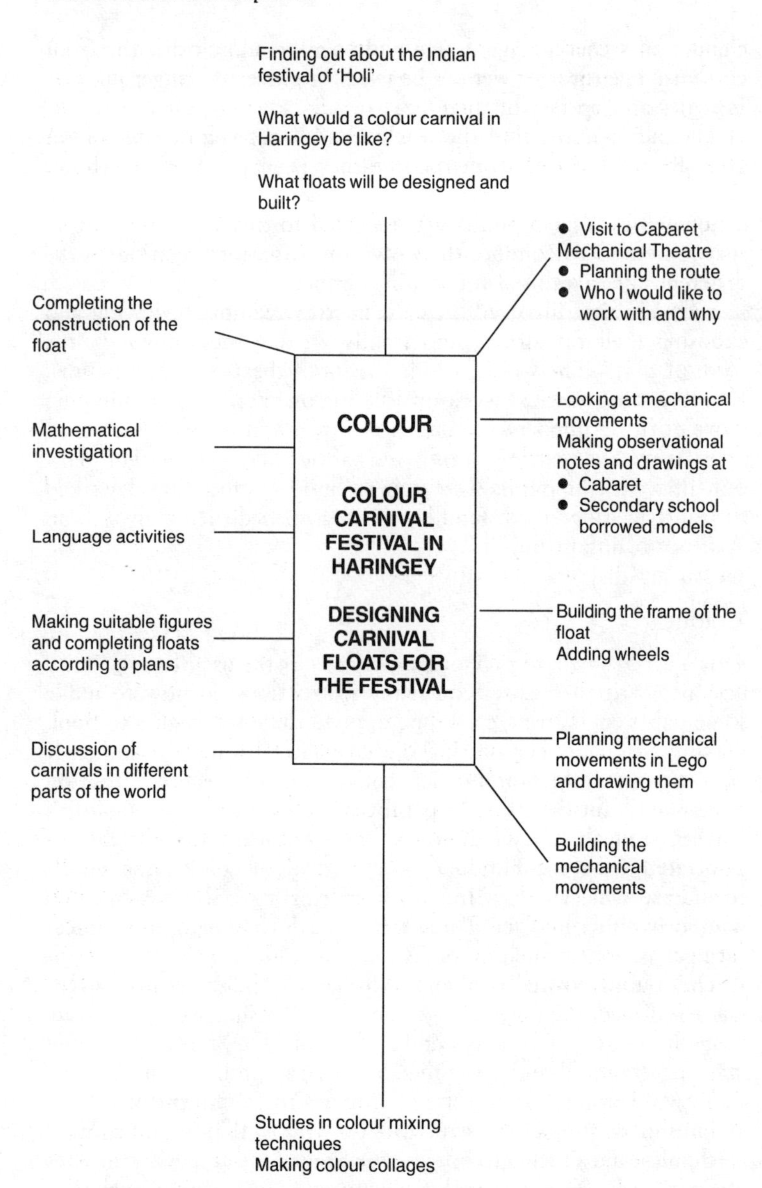

Figure 7 Breakdown of classroom activities planned for a topic on 'Colour'

methods for making gears, cams and pulleys.

Other language activities specific to this project included mapping out the journey, reporting on the day out, writing down who they would like to work with and why, and filling in a log of the work done on the project and the progress of it, including decisions, mistakes and how they felt about the work. They studied and experimented with colour mixing, colour pigmentation and tone, as well as engaging in other work associated with carnivals, festivals, dance and music.

'Transport and machines'

This class were introduced to tools by constructing a rectangular base as a chassis and attaching wheels (as described in chapter 3). These were then used as bases to construct some form of moving machine. The pupils experimented with wheels and then proceeded to test the variety of methods they had devised for attaching wheels and axles on the floor and on ramps. The latter proved useful for measuring distance and time and finally introducing the idea of speed.

Having tested their chassis and made some decisions on efficiency, they proceeded to build these up into a series of machines they had chosen as part of their topic (cranes, vehicles, lorries, buses). Resource books were useful to give ideas about construction and function of the different machine parts.

It is possible to develop this project in a number of ways; for example, as an energy project, whereby the pupils are given the further problem of propelling their machines a certain distance or up a ramp using a number of sources of power (air, water, elastic band power, electricity). (It is possible also to use clockwork wind-up mechanisms and cheap electric motors.) On the other hand, interesting work on journeys, terrains and adventures across different lands can be used as a basis for a project similar to that described in chapter 5.

'Boxes'

A number of classes made simple boxes as containers, studying specific aspects of these. The major technical problems shared were how to fit the lids on and what types of lids they wished to design. The choices are numerous, and many boxes were brought in to

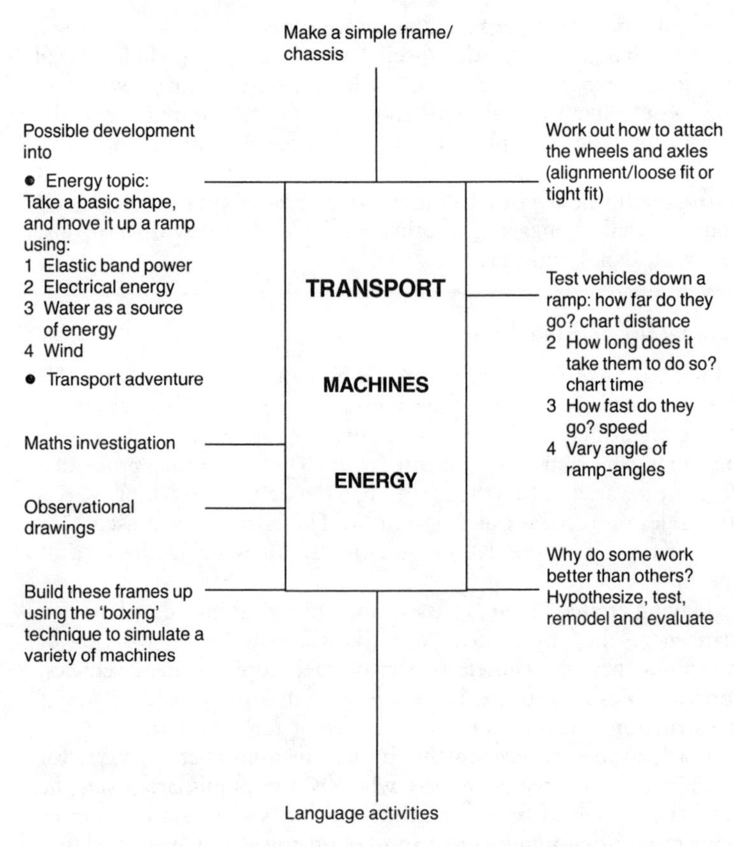

Figure 8 Breakdown of classroom activities planned for a topic on 'Machines'

study: removable lids; hinging (a variety of these exist); pivoting lids; sliding lids. It required very careful observation of hinges to understand how they worked and then construct them from available materials.

A study of boxes and their functions provided interesting discussions. Boxes and their secrets were used as the basis of story writing activities, as well as existing stories about boxes ('Pandora's Box', 'Bugs in a Box') that were used as starting-points for imaginative work. In a way similar to the 'growth' topic, pupils could use stories such as 'Where's Spot?' and 'The Haunted House'

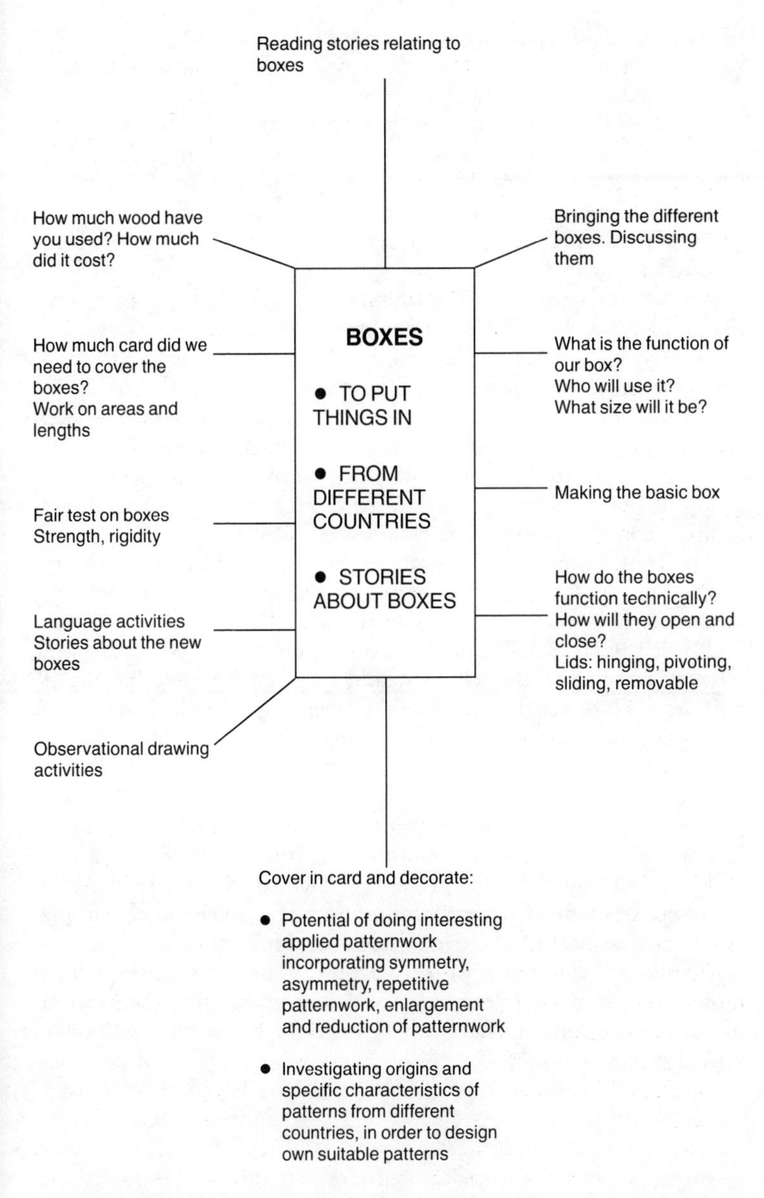

Figure 9 Breakdown of classroom activities planned for a topic on 'Boxes'

to have things popping out of different structures. One class made bird boxes and had to pay special attention to waterproof glues and the choice of materials used to cover them.

By carefully removing the electronic parts of musical cards (available from most stationers), and placing them in the lids and sides of boxes, it is possible to make musical boxes.

The different sizes of the boxes provided many additional mathematical activities, including measuring the area of card required to cover the box.

A study of patternwork and symbolism was used as a means of designing boxes from different countries around the world. Islamic patternwork, African mask designs, Egyptian hieroglyphics and Japanese designs were all used as a stimulus for the study of pattern, symmetry, asymmetry, enlarging and reducing these, repeating them and finally applying the designs to cover the boxes. Thin veneer cut to the shapes and stuck on to card was used by one group of pupils in another school to produce very rich and sturdy looking containers that were finally varnished.

The school experiment at Down Lane Junior School proved very successful. It brought the pupils together, and many classes described their work to the rest of the school at assemblies ('Show and Tell'). Classes visited others to see how the work progressed. Visitors arriving to the school would be given full descriptions of what each class was doing before they went into the class, and there was a constant supply of pupils keen to describe their work and show it off.

It proved a positive way of bringing the staff together in a practical way, to share the problems, frustrations, and, most importantly, the successes. It built up the confidence of the staff to tackle more open-ended problems and to make design and technology activities a more familiar part of the general class topic and the tools part of the overall class equipment.

The second case-study in this chapter is that of a school which developed the work over a period of two years, and developed a method for recording the work in order to build up a workable school policy.

Design and technology as a focus for curriculum development

School: South Haringey Junior School
Headteacher: Mr James Shepherd

At South Haringey Junior School, curriculum development in design and technology was seen as central to topic work in the classroom, and every topic would have practical problem-solving activities as a major feature of the topic work, giving rise to work in all other areas of the curriculum. We describe the developments over the two-year period.

The first year

The headteacher had planned for design and technology to be the major focus of development in the school, thus devoting a considerable amount of the school's capitation to initiating and resourcing the school with necessary tools and materials.

Four members of staff were encouraged to attend in-service training aimed at developing their confidence in handling the tools. These teachers were located because they had different interests and areas of responsibility: science and technology, art and design, language, English as a second language. They also taught Year 3 and 4 classes in the school, and could therefore start the work with the youngest pupils to enable progression to be mapped in later years as the work spread throughout the school. The first year of curriculum development was spent developing the following:

1 The experience and confidence of the teachers concerned, so that they might assist others in the future.
2 Building a stock of tools, resources and materials for the future.
3 Ensuring these activities were a focus for general curriculum development and central to the topic work in the classrooms.
4 An exhibition of the work at the end of the year would be used as a means of encouraging other teachers and explaining the work to parents.
5 The teachers concerned would bring their different sets of expertise and interest to bear on their particular projects, in order to show breadth of possibilities across the different areas of the curriculum.

6 Studying the effects of these activities on the learning developments of the pupils.

The work by the end of the year was impressive. The teachers had introduced pupils to the tools and developed the work further. They had used a number of ideas from LEA exhibitions and adapted them to their own requirements, topics and interests. The work in the lower years had spread to the older classes through the enthusiasm of the pupils.

The teachers were able to develop particular aspects of the work in their classrooms and study the links of design and technology to their particular curriculum responsibility, focusing on their interests and particular expertise to enable the pupils to produce work of exceptional quality, demonstrating the potential that is capable of being produced by the whole ability range of a class. Many of the ideas for activities suggested throughout this book were developed during this intensive process.

The second year

The main objectives for the second year of development in the school could be listed:

1 To spread design and technology in all the classes and involving all staff.
2 To reflect on the work of the previous year, record it and build upon it.
3 To study the place of design and technology activities in the primary curriculum and classroom.
4 To develop a workable and practical school policy for design and technology.

Further tools and resources were brought in, so that every teacher had easy access to a simple tool kit. Other teachers were encouraged to attend courses, and workshops were held in the school. Each of the teachers with previous experience was teamed up with other teachers and spread throughout the age groups. One teacher was able to work alongside teachers in their classrooms, with the support of the headteacher.

In order to develop a policy, it was important to start to record development to date and to generalize from the previous year's experiences. The school had adopted a GRIDS (Guidelines for Review and Internal Development in Schools) process to monitor and plan this development.

The principles of this process are based on a collective development and sharing of ideas to develop the work throughout its stages. The stages involved are:

1 Decide where the school is now.
2 Decide where the school wants to go.
3 Decide how it will get there.

The first stage of the process involved the whole staff, working together to record the work in the previous year, and to monitor ongoing developments. In order to have some form of structure, it was decided to break up into four working parties. The staff would subdivide into three sub-groups: Language, Creative and Aesthetic, Science and Technology. The whole staff would work together on the mathematics review.

Each of these groups would be responsible for gathering all the information particular to their area. The aim was to ask every member of staff what particular language, science, mathematical or other activities had arisen from the projects they had been involved in. These would be described and placed in context. The skills, concepts and attitudes developed would be recorded. Each working party would collate the information, and then report to a whole staff meeting for the purpose of discussion and further modification. This process was a lengthy one, taking two terms to complete. It revealed possible common activities that could be developed throughout the school and showed the potential of the work as a cross-curricular tool. This period also included a review of the materials and resources the school was using, and how they were being used and accessed by staff and pupils.

The second stage was for the staff to decide how they hoped to develop the work and in what direction they would be developing. At this stage the staff carried out a review of the books, posters, videos and other resources available on the market, as well as recent relevant publications. Discussions followed on how these might be used to develop a policy for the continuation and development of the work.

The staff developed a series of questionnaires that they would circulate among themselves to be used as a checklist to ensure agreement, continuity and progression throughout the school and to list the following:

- The processes involved in design and technology
- The materials used
- The techniques and practical skills covered
- The areas of technological knowledge arising out of the problem-solving process

Topic planning was central to the work in South Haringey Junior School. At the time of going to print, some work had been done in this area, and this is recorded, in part, in chapter 8. The work has since been used and developed in other schools.

The areas of discussion, the processes and pointers for the development of a school policy are included in the Appendix. Separate sheets are included on the types of materials that may be used in design and technology activities, and the sorts of questions a school staff may wish to address when developing a policy for use and resourcing of materials.

The second stage of this GRIDS review also involved important decision making regarding where the next step for the development would be.

The school had initiated change and development using design and technology activities as a focus for a wide-reaching cross-curricular review. The aim was to carry on with the cycle of in-service, classroom implementation, discussion and review, but to narrow and focus future development on specific curriculum areas (the next, in their case, being science and technology). The policies and methodologies that the school had developed through a study of design and technology across the curriculum provided a guiding basis for the individual curriculum areas. In future years, mathematics, language and other curricular areas would be reviewed.

This chapter has shown how it is possible for a school to focus on an area for curriculum development in a collective manner. The process of co-operation strengthens the impact of the development on pupils and staff. Monitoring and recording the changes and their effects can be a way of developing a coherent workable policy for the particular school. It has not been our intention to record the specific policy outcomes, but to concentrate on the method used to arrive at them, and to provide pointers that teachers in other schools may find useful while developing their own approaches.

In both cases, involvement of pupils, staff and parents was important to the success of the work. The tripartite relationship ensured the continuity of the work. (At Down Lane Junior School, the workshop for a hundred parents was particularly successful, since it used the pupils to organize the evening and activities.)

The two schools have also been chosen because they both decided to use design and technology as a means for developing two different curricular foci: language, and cross-curricular topic work. In both cases, curriculum development was planned to a realistic timescale, to enable change to happen and then reflection on the nature of that change.

Throughout the book, we have argued that the methodology and processes of design and technology activities are essential in their own right as worthwhile educational activities that develop pupils' problem-solving capabilities and their understanding of their environment, and give them a sense of being able to effect change. We also argue by means of case-studies that these design and technology activities are cross-curricular in nature. However, it is important that the activities are well planned, organized and resourced to ensure the success of such an integrated topic approach. The next and final chapter provides a framework to help organize topic work.

8

Topic planning

In chapters 2–7 we have shown, through examples of work, how different teachers have tackled design and technology activities in their classrooms with different ages of children and in a variety of ways. Throughout, the work has always been incorporated into a topic. The projects have been used to engage pupils in technical and investigative problem solving, and often include open-ended problem solving. The projects and activities have also been used to address language, science, art, human and social areas of the curriculum.

However, we have not shown any of the detailed preparation that often accompanied these activities. In most cases, the teachers were also tackling the use of tools in the classroom for the first time. Organization and planning was of paramount importance to the success of each venture.

In chapter 4 we described in detail the activities of a Year 4 class as part of their topic 'Masks'. In this chapter we show how those activities fitted into an overall scheme of work and the type of planning that preceded the work in the classroom. We use this method of planning to develop a model that can be used and adapted by teachers in general.

After brainstorming the idea of masks and planning a wide range of activities, the teacher decided to focus on a more restricted scheme of work. This scheme of work is recorded as a series of sequenced projects (see Figure 11). Each of these projects included a series of tasks that were planned to draw out different skills, concepts and attitudes particular to areas of the curriculum. The teacher listed these and ticked off the appropriate areas (two of these sheets are shown as examples in Figure 12).

When the teacher was asked why she had completed this breakdown, her answer was that she would be able to look at her 'ticks' and have an overall view of what she had missed out. She would then plan her next topic to have greater focus on those areas she felt the pupils had not experienced to the same depth that she hoped for.

The National Curriculum focuses teachers' attentions on what they teach in terms of breadth, balance and appropriateness. The choice of relevant work that draws on the different attainment targets in the statutory orders is essential. Some teachers may argue that the National Curriculum documents force them to teach separate subjects so that they can 'cover' the requirements. The topic approach remains, however, the most effective teaching and learning strategy that draws on the different areas of knowledge, because it provides pupils with the links that give meaning to the diverse areas of knowledge that we call 'subjects'. Using the topic can also ensure breadth and balance incorporating core and foundation areas.

One of the greatest concerns of teachers is the method of recording and monitoring work to ensure that attainment targets are being addressed in a systematic way. There are excellent methods of monitoring skills and concepts in each of the core areas. These are all very useful for a school that is focusing on one curriculum area and studying that area in depth for the purposes of staff and curriculum development. However, there is a need for models which address all areas of the curriculum across the topic work. These methods need to be easy to use, the least time-consuming possible and helpful for planning, recording and monitoring achievement.

We offer the reader one model that has been used as part of the topic approach. It starts with web diagrams of ideas and develops them to a detailed scheme of activity of teachers' plans. This is translated into classroom charts to monitor the work. Some schools have developed this further by recording outcomes in a pupil's individual record of achievement or profile.

The main advantage is that the teacher can plan a detailed series of activities, and use these to highlight the attainment targets that are going to be tackled. This gives a clear picture of what is missing, so that these areas can be incorporated in another topic or additional activities that focus on those attainment targets.

There is a growing trend, reflected in changes in teacher training, for teachers to have expertise in particular curriculum areas. Curricular responsibilities are shared among the staff of the primary school. The school consists of a team of teachers who

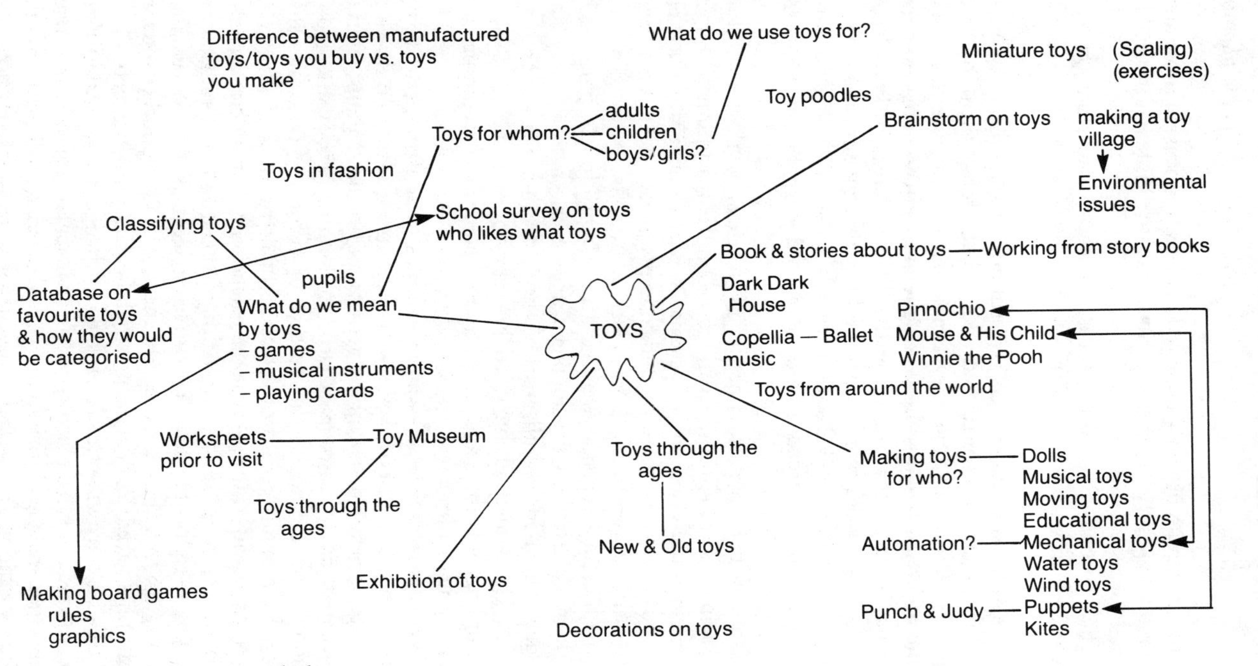

Figure 10 Brainstorming of ideas

work in a common way in their classrooms, but use team members to make that work more effective. Consequently there is a need for teachers in schools to negotiate and agree on common approaches to address the requirements of the National Curriculum. It is no longer possible to work in an isolated manner in a classroom, without reference to previous work or other colleagues' expertise.

The following suggestion for a school structure to approach topic planning has been developed from the work of teachers individually and in a number of schools that have decided to take on the task in a collaborative manner. Figure 13 condenses the following five stages in a diagram.

Stage 1: Working in groups teachers brainstorm a particular choice of topic. Brainstorming is a useful technique, because it gives the teacher a number of very divergent ideas and options for work from different viewpoints. Each teacher will have a topic brainstormed in this manner. Figure 10 shows an example. It is useful to adhere to the rules of brainstorming to make this method work. These are:

1 all suggestions are written down
2 no criticisms of any suggestion
3 the time must be limited (no more than 5 minutes)
4 every comment must be recorded

Stage 2: The next stage is to extrapolate those ideas that are practicable, manageable, and appropriate to the class and its particular circumstances. Thus the teacher can plan out a generalized strategy and sequence of broad activities to follow (as illustrated in Figure 11 or any of the outlines in chapter 7). The teacher could do this with other colleagues or independently.

Stage 3: This stage involves more detailed planning of the overall ideas decided upon. Following the broad outline of topic-related activities, it is possible to make a further in-depth study of the specific areas of learning that are being tackled. The broad plans have to be broken down into more detailed classroom activities. At this stage, it is useful to discuss the plans with teachers in the school who have a particular expertise or are responsible for a specific area of the curriculum for their suggestions (e.g. additions, resourcing materials, support with producing worksheets), in order to draw out those particular concepts, skills and knowledge that

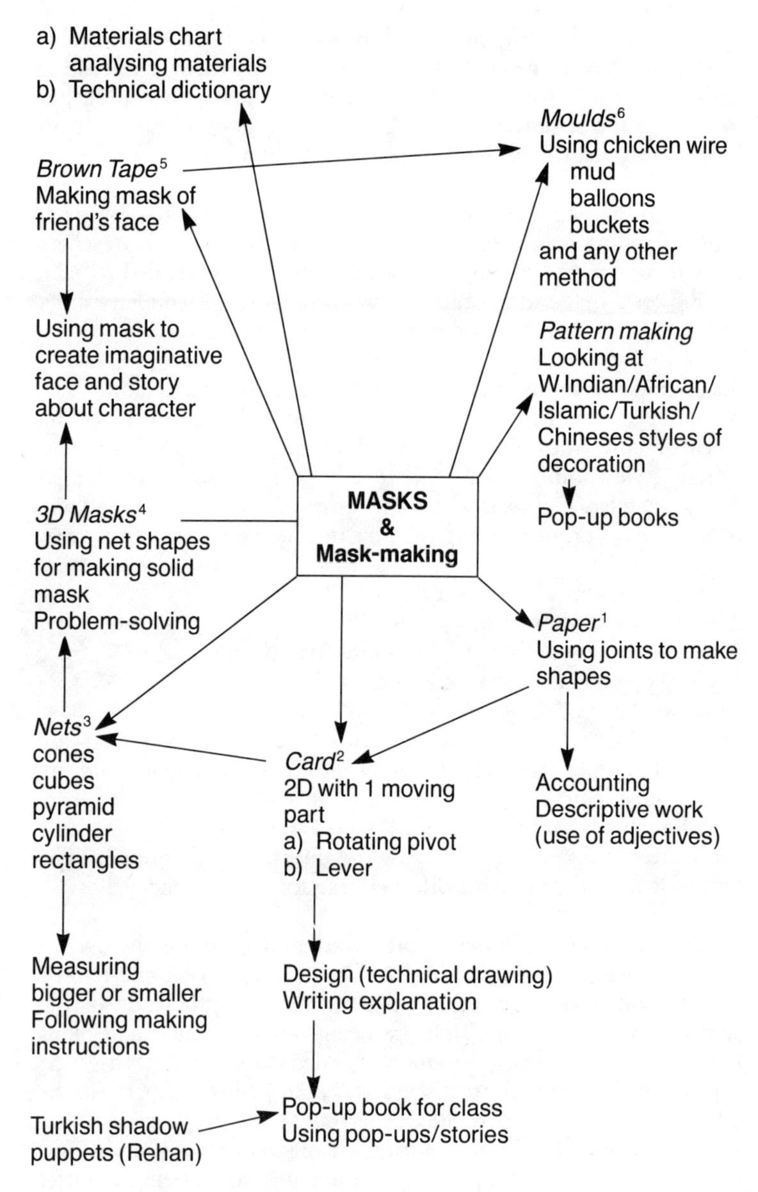

Figure 11 Outline of scheme of work (masks)

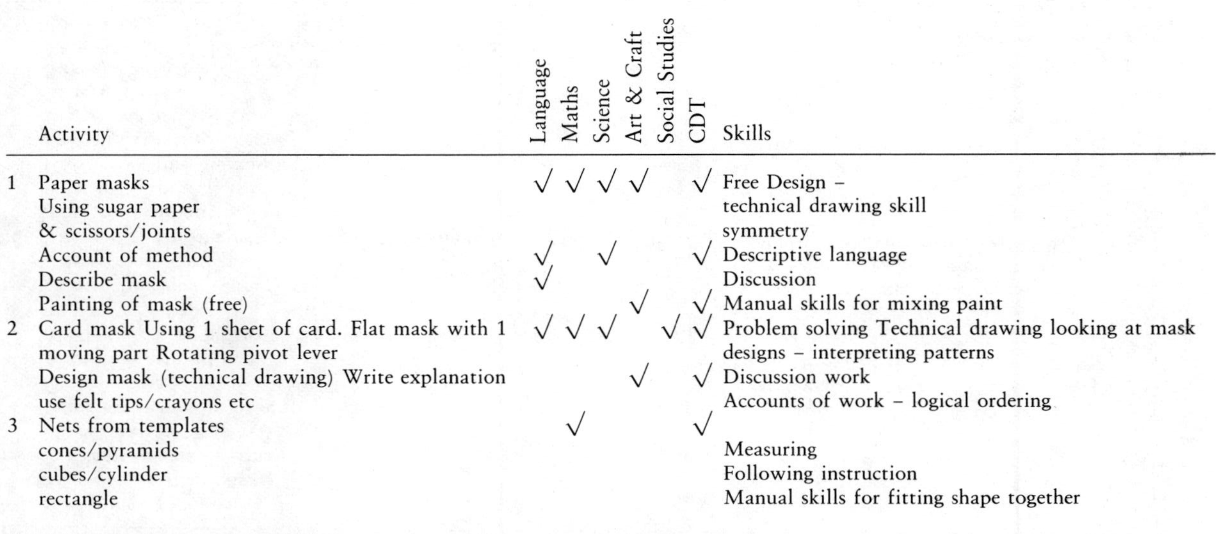

	Activity	Language	Maths	Science	Art & Craft	Social Studies	CDT	Skills
1	Paper masks	√	√	√	√		√	Free Design –
	Using sugar paper							technical drawing skill
	& scissors/joints							symmetry
	Account of method	√		√			√	Descriptive language
	Describe mask	√						Discussion
	Painting of mask (free)				√		√	Manual skills for mixing paint
2	Card mask Using 1 sheet of card. Flat mask with 1	√	√	√		√	√	Problem solving Technical drawing looking at mask
	moving part Rotating pivot lever							designs – interpreting patterns
	Design mask (technical drawing) Write explanation				√		√	Discussion work
	use felt tips/crayons etc							Accounts of work – logical ordering
3	Nets from templates		√				√	
	cones/pyramids							Measuring
	cubes/cylinder							Following instruction
	rectangle							Manual skills for fitting shape together

Figure 12 Breakdown of curriculum areas from scheme of work (masks)

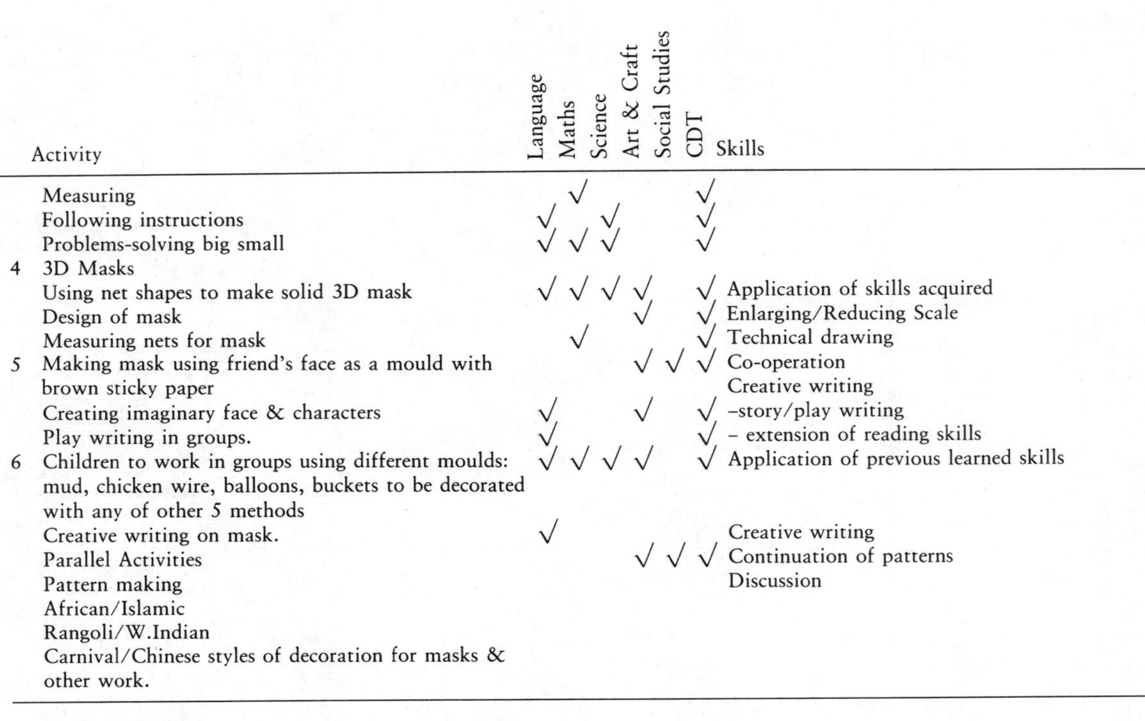

	Activity	Language	Maths	Science	Art & Craft	Social Studies	CDT	Skills
	Measuring		√				√	
	Following instructions	√		√			√	
	Problems-solving big small	√	√	√			√	
4	3D Masks							
	Using net shapes to make solid 3D mask	√	√	√	√		√	Application of skills acquired
	Design of mask				√		√	Enlarging/Reducing Scale
	Measuring nets for mask		√				√	Technical drawing
5	Making mask using friend's face as a mould with				√	√	√	Co-operation
	brown sticky paper							Creative writing
	Creating imaginary face & characters	√			√		√	–story/play writing
	Play writing in groups.	√					√	– extension of reading skills
6	Children to work in groups using different moulds: mud, chicken wire, balloons, buckets to be decorated with any of other 5 methods	√	√	√	√		√	Application of previous learned skills
	Creative writing on mask.	√						Creative writing
	Parallel Activities				√	√	√	Continuation of patterns
	Pattern making							Discussion
	African/Islamic							
	Rangoli/W.Indian							
	Carnival/Chinese styles of decoration for masks & other work.							

Figure 12 (continued) Breakdown of curriculum areas from scheme of work (masks)

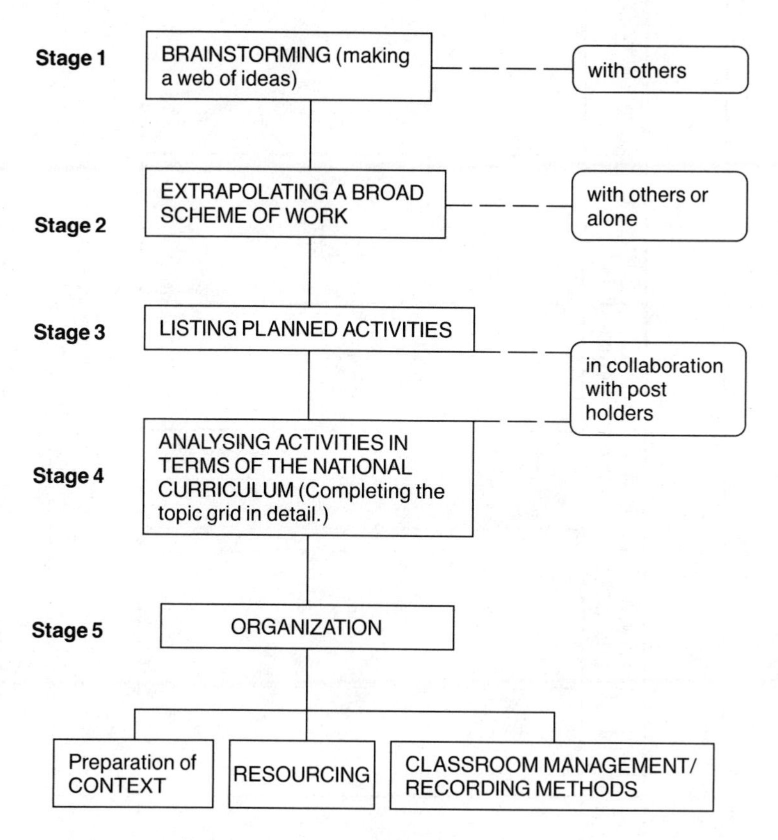

Figure 13 School strategy for topic planning

can be highlighted in the activity. It is possible to use the whole school's expertise to address all areas of the curriculum. The teachers can match the attainment targets of the National Curriculum with the detailed descriptions of the activities and those areas that they wish to highlight.

Stage 4: At this stage, it is possible for teachers to fill in a detailed grid of their topic planning in terms of lists of anticipated classroom activities, and how each of these may address one or more of the attainment targets in the National Curriculum. The

TOPIC				CONTEXT OF THE WORK					
Describe activities in order of plan	LANGUAGE			MATHS	SCIENCE & TECHNOLOGY	HUMAN & SOCIAL	CREATIVE & AESTHETIC	DES & TECH	RESOURCING
	S&L	READ	WRITE						

Figure 14 Topic planning grid

grid shown in Figure 14, has been designed over a period of time, and after many trials for this purpose. Teachers have enlarged it to A3 for ease of use. It has also been adapted to suit the school's particular circumstances. In the terms of the grid, 'Context' answers questions such as:

- Why choose this topic?
- How does it apply to the pupils' world?
- Can it lead to a broadening of pupils' experiences by widening their perspectives?
- What learning outcomes can be gained from it?

This enables the teacher to put the work in a wider perspective so that he or she may use the pupils' own experiences as a starting-point and then widen their horizons by making them aware of related issues outside their immediate surroundings. Preparation of the context for the topic enables the teacher to highlight a number of the human, social and moral areas of the topic. Many of the issues raised in chapter 1 can be addressed by the teacher at this stage.

Teachers can record separate language activities: speaking and listening (S & L), reading ('Read') and writing ('Write'). This enables teachers to differentiate and record the following speaking and listening activities:

- between pairs of pupils
- small groups of pupils
- pupil with adult
- small group/large group with adult.

The activities listed on the left (such as verbal report to class on progress, working in groups of five to discuss a problem, reading a particular book) describe the particular circumstances for the differentiated language work. Teachers can list the particular learning outcomes they hope to achieve from the particular reading or writing activity as well (completed log of recorded work, sequencing instructions, project evaluation, completed cutting list, or evidence of research, notes from non-fiction reading).

'Human and social' refers to the areas of learning associated with social studies, history and geography. The 'creative and aesthetic' areas of the curriculum include the performing and

List of activities	Areas of learning & Attainment Targets involved											Elements of learning
Describe activities in order of plan	*Maths*	*Aesthetics & Creative*	*Science & Technological*	*Human & Social*	*Physical*	*Language* *speak* *1*	*listen* *2*	*read* *3*	*write* *4*	*Context*	*Resourcing*	
Marbelling in blues to make cover of books		✓		Appreciation of other children's efforts								Concepts & Skills Knowledge Attitudes
Make folders	AT1 AT2 AT8 AT3 Applying number	✓					✓					
Reading poems & stories								Express opinions Reasons for preferences Discuss poems Develop tastes Infer and predict from text 4a 4b 4c 4f 5a,b,c				
Writing poems Suitable ??? poems		✓							4a, b, c, d, e, f, g, h, i			
Water stories		✓							4a, b, c, d, e, f, g, h, i			

Figure 15 An example of a curriculum analysis

expressive arts (e.g. art, music, dance and drama). A separate column has been added for design and technology, so that teachers can tick off appropriate programmes of study covered by the activities.

Stage 5: The following method is particularly successful when it is planned one term in advance, thereby allowing the teacher to prepare well in advance for any trips, outside visitors, books, resources, or personal reading-up to familiarize themselves with a particular aspect they are not sure of. The fifth stage would enable the teacher to:

1 Find and collect resources. The importance of these resources cannot be overstated. Teachers can look for appropriate reading: fiction or poems that relate to a topic can be found, as well as non-fiction source materials for open-ended problem solving. The choice of materials will also enable teachers to highlight a worldwide perspective to their work. Examples of different cultural representations of dragons (Chinese, Welsh, George and the Dragon) can greatly enrich pupils' images.
2 Plan any trips, visitors or videos that will stimulate the topic and bring the work into a context of some relevance to the pupils.
3 Enable the teacher to prepare for the topic in terms of reading, any necessary skills they wish to develop.
4 Decide and prepare for the different forms of classroom organization and recording necessary.

While this general topic planning model has taken into account methods for planning classroom work, it is also important to devise instruments for monitoring the progress of the plans in the classroom. Figure 16 shows an example developed and used by teachers from North Haringey Junior School. The teachers have used this grid to ensure that the model proposed has been adapted to suit the schools' particular purposes. Teachers have adopted the topic grid for their own planning purposes. In many cases, teachers have decided to adopt the model in a series of stages:

1 Use it initially as a tool to reflect on what has already happened in the classroom, thereby getting used to filling it in.

2 Fill it in without worrying about attainment targets but as a planning tool.
3 Add the necessary attainment targets at a later stage, and reflect on additional activities that would address additional ones. This last stage was useful for teachers to familiarize themselves with the language of the statutory orders.

The overall approach to topic planning outlined in this chapter is not new to teachers and schools. It is useful to consider the advantages of such an approach:

1 When used as a whole school, it encourages a team approach to curriculum development. It gives teachers opportunities to talk about their topics together, and discuss the reality of issues of continuity and progression.
2 It makes more effective use of the expertise available across the school and decreases individual teachers' sense of isolation.
3 Planning topic work in advance gives teachers the time to prepare more effectively and find the necessary expertise and resources they may require from outside the school.
4 It offers teachers methods of planning, classroom management and recording that are linked to each other, so that these demands are not seen as separate.

However, there are disadvantages as well, especially when teachers use the grids as a package, without discussing how this best suits their particular circumstances and adapting it as necessary. The main advantages are that:

1 The teacher keeps an easy-to-read and useful form of monitoring progress of the general plan of the class as a whole.
2 The teacher can monitor individual pupils' progress throughout the work.
3 The teacher can record pupils' contributions to discussions to ensure an even distribution of opportunity.
4 Any instruments used for monitoring can be easily extended to incorporate the teacher's assessment of levels of achievement. Furthermore, all these records must be easy to use and not seen as separate from the planning process.

The activities listed in the topic planning grid can be transferred to

/ started ∧ half way △ completed	List of activities														
NAMES	Target ATs														

Figure 16 Monitoring grid

the top of this grid. Pupils fill in the chart, following consultation with the teacher. The triangle could be extended to a square. The pupil starts the shape when the activity is first started. If a piece of written work is listed, as an example, the teacher may decide to ask pupils to draft and redraft. At an appropriate stage, the teacher will ask the pupil to complete the second side. When the written piece is complete, the shape is finished off. Attainment targets can be added at the top of the grid under each activity.

When the grid is complete, it is removed and teachers can insert the levels of attainment achieved (where they wish to do so). In this way teachers can see, at a glance, not only how the work is progressing, but also the levels which pupils have reached and the particular activities that were used to assess the level of attainment (that particular piece of work can be kept for future reference, if possible). This information can be transferred to some form of pupil profiling. The added advantage of such a grid is that pupils are aware of the range of work they will be engaged in and can also monitor their own progress.

The topic planning and classroom management grids can also be used to plan and monitor non-topic-related activities.

Key Stage 1 teachers have adapted this grid by using symbols in the place of written activities (e.g. a book for reading, an ear and mouth for carpet discussions, numbers for mathematical activities, and others).

Throughout the book we have shown examples of how teachers have used a theme or topic as a means of focusing on a particular design and technology project. The definition of this project and the activities that are associated with the practical project, such as report writing or solving technical problems, engage pupils in practical applications of science, mathematics, language and art skills and concepts. In chapter 7, we described the process by which two schools focused on design and technology development as a whole school. The chapter highlighted the importance of developing a spirit of teamwork and a common sense of direction. Both schools had identified a series of stages to enable curriculum development to occur. In this chapter, we emphasize the necessity for teamwork in helping teachers develop their planning work in individual classrooms. The model for topic planning uses teachers' expertise across the school to ensure that the activities that are planned in relation to design and technology projects can be used more effectively to focus on other areas of the curriculum. We hope that, through these examples and methods of recording, teachers will continue to work in the integrated manner that has proved so successful in the past.

Conclusion

'Sharon never talks about what she does in school – now all I hear is Design and Technology this and Design and Technology that. Do they do anything else in school, and what is this Design and Technology?'

This parent's remarks highlighted fundamental questions about design and technology in the primary school classroom.

- What is design and technology?
- Why are the children doing it?
- Where does it fit into the curriculum?
- What about the other areas of the curriculum?

We have used a series of case-studies to describe what we mean by design and technology in primary schools. Through a series of statements we have attempted to clarify what is most important about these activities. This can be summarized as follows.

Design and technology is concerned with problem solving generally. The wider the scope of the problems, and the more open-ended they are, the more design and technology activities are possible. If they concentrate on how things work or are made, they are more technological. In nature, open-ended problems respond to needs. The problems are set within a framework that pupils can relate to. Their experiences and the stories they read can be used as starting-points for these activities.

Pupils are engaged in these activities to gain control over their environment, to improve their environment, to make sense of the world in which they live, and actively to participate in it. In specific educational terms, design and technology activities give a focus for the development of other areas of the curriculum; for example, as a focus for language recording as shown in chapter 7. Design and technology problem-solving activities require pupils to communicate – to show what has been done, and to communicate instructions, results, ideas and understanding to others.

Design and technology can integrate and focus the curriculum as it draws on other areas of knowledge. This was seen very clearly in chapters 4 and 5, where mathematical and scientific knowledge and processes were drawn out of mask making (nets), an environment project (measuring and scaling), making mechanical toys (mechanisms), a transport problem (problem solving) and a water project (materials and sinking and floating).

A further question is asked by those who are new to design and technology activities, and that is 'how do we introduce it?' Each classroom teacher whose work is described in this book has drawn on their knowledge of related activities. We have looked at the concerns and needs of those teachers to provide a framework within which design and technology can be embarked upon with excitement, not apprehension.

What seems to be crucial to the success of these activities in the primary classroom is the use of a topic approach, so that the context has meaning for pupils and teachers alike. In chapter 7 we related the experience of two schools in introducing design and technology through this approach. The topic approach is also supported by the recommendations of the Design and Technology Working Group for the National Curriculum, and in chapter 8 we offer a model for introducing design and technology within a topic. Also in this chapter we have shown how attainment targets and programmes of study, across the core subjects, can be planned for, monitored, recorded and achieved.

The implications for design and technology in the National Curriculum are various. In-service training for classroom teachers, facilities and equipment, provision of hardware for information technology, implications for equal opportunities – these are issues which will be addressed over the years to come. Initially, however, teachers will be using and developing skills and expertise they have been so successful with in the past to embrace, integrate and enhance design and technology in their classrooms. We hope that this book will support and encourage teachers to participate in the development of this exciting area of the primary curriculum.

Appendix

Resources

In this appendix we have brought together a number of resource materials which have proved useful to teachers working with us. What to use and where to find it are common questions asked, and in the first section we have attempted to answer these. Suggestions for identifying and classifying materials and their uses are considered at the end of this section.

Tools and materials

Tools and materials for design and technology activities are many and various. Many are already to be found in the primary classroom. These include paper, card, tissue paper, balsa wood, crayons, paints, scissors, rulers, tape-measures, paint-brushes, glue-spreaders, hole-punches and treasury tags. Much of the material needed for construction can be obtained through collecting cereal boxes, plastic containers, egg boxes and so on. Resistant materials, such as wood, metal, and plastics, can be obtained through scrap schemes as well as local education authority (LEA) stockists. Many classrooms have access to workbenches and tool sets for construction work; provision for these activities will vary from school to school. Below are listed items of equipment for working with resistant materials, as these may be the most recent additions to the primary classroom.

- Vices: There are many types of vice available. These can be attached to any table edge. The teacher should always ensure they are firmly on before work commences.
- Bench hooks: Recommended for use with older pupils as one hand is needed to hold the wood against the block.
- Jigs: These make cutting and sawing easier and fit into a vice with slots to help guide the saw blade. Good if only one section of wood is to be used.
- Junior hacksaws: These saws are designed for cutting metal. However,

they are excellent for use at primary level for cutting all types of resistant materials, such as timber, plastics and metals There are many different types, and different LEAs will recommend certain types whilst positively discouraging others. Generally, it is safer to have one with an enclosed handle so that children are not in danger of hitting their knuckles.

- Abraders: These are useful tools for sanding the ends of cut timber and for simple shaping. The varieties of abraders available are increasing. They may seem expensive, but are worth the cost in the long run.
- Hand-drills and drill bits: These can be costly, some brands especially so. There are some especially designed for primary use. However, it is important to check that they are sturdy enough and will withstand a lot of use. Drill bits should be high-speed steel (HSS) and a variety of sizes are advisable (3, 4, 5 and 6 mm).
- Bradawl: This is useful when drilling to start the hole. Use the type with a flattened point.
- Hammer: It is best to purchase lightweight hammers (2.5 oz Warrington hammer). Do not use toy hammers.
- Try-square: This is unnecessary for young pupils, but it may prove useful for teachers to mark timber accurately. It is a tool used to mark a line at right angles across a piece of timber, as well as checking if a sawn piece has been sawn squarely.
- Pliers: Used for bending wire, as well as cutting thin wire.
- Straight edges: These are used for drawing straight lines on timber. Rulers would suffice.
- Electrician's screwdrivers: These are used for attaching wires on to bulb holders.

A class set of tools

The following set of tools represents a basic set used by many schools starting out on designing and making activities:

5 junior hacksaws and spare blades
1 hammer
drill bits (3,4,5, 6 mm plus several 2 mm)
3 abraders
metal safety rules
nylon cutting mats
5 table vices
1 hand-drill
3 try-squares
1 bradawl
craft knives

Where to buy tools

Several education authorities provide their own service for the purchase of tools, with particular items for use in primary schools. Check with your local authority/local teachers' centre. Increasingly, general tool suppliers are producing sets of tools suitable for primary work.

Additional tools and materials

The Middlesex Electricity Kit 5–13 ('Learning about Electricity by Total Discovery' 35) from Middlesex Polytechnic, Trent Park, London.

pliers	cutters
wire strippers	electrician's screwdrivers
small bulb holders	stranded core single connecting wire
motors	buzzers
bulb holders	bulbs (3 v, 6 v, 8 v)
crocodile clips	batteries (4.5 v, 9 v)

Other useful tools and materials include:

welding rod
cotton reels
dowel rod (4 mm, 6 mm)
plastic wheels
wooden wheels
cardboard wheels
reels of polythene sheet

wheels (tin cans, squeezy bottles cut in half and pushed into one half – (two tops provide an instant axle, two bases can be drilled through).
garden cane and rubber bands together can be built into triangular structures; add wheels and plastic sheet to make land yachts.

syringes and tubing with connectors
propellers
glues – PVA, wood glue, paper and card glue
glue gun – not for small children to use but for teacher use
araldite

wood – scrap pieces from timber merchants
square section wood – 9 × 9 mm
lollypop sticks

corrugated plastic sheet

fabric pieces	leather scraps
corks	screw eyes
balloons	pulleys
washers	beads
marbles	art straws

string
staplers
paper clips
thread
paper fasteners
parcel tape

soldering iron – for use with third year juniors upwards

squeezy bottles
plastic containers
paper cups
water containers (plastic)
yoghurt pots
paper plates
plastic spoons

Where to look for support, information and materials

membership of scrap projects – tremendous value
surplus buying agencies
local authority teachers centres – for information
local polytechnic and further education colleges
technical suppliers
local secondary schools – D & T, Art departments

SATRO'S (Science and Technology Regional Organizations)
Association for Science Education
National Association for Design Education
Design Council
Design and Technology Association

Questions to consider about the use of materials in schools

1 What materials are used in the classrooms? What type of activities have you used them for? Make a list as a record.
2 How are they sorted, categorized and stored?
3 Who has access to them? Do the pupils have access?
4 How often are they used?
5 How are these materials used? Are pupils shown specific techniques for using them, (e.g. slab, coil or thumb pot techniques in clay work)?
6 Do the class generally use the material on the same project (i.e. all make a coil pot)?
   Does every pupil have the opportunity to handle the material? How is this monitored?
8 List the types of techniques you associate with each of the different materials.
9 Are the materials available for use on other projects where the teacher does not anticipate their use? If so, how are the materials made accessible to the pupils?

10 Have pupils been involved in testing the materials? – their properties, such as malleability, strength, rigidity and structure etc., their source and classification, the nature of any changes in water at different temperatures etc.
11 Have the pupils been involved in testing various adhesives appropriate to the respective materials?
12 Are the pupils aware of the different methods of working the materials (e.g. joining, cutting, shaping and fabricating techniques)?
13 Are they aware of how these materials are used in other parts of the world?
14 What are the ecological implications of the material?
15 Are there any historical points of interest about the material? Are there any particularly interesting uses of the materials in different parts of the world?

Figure 17 shows the different types of materials that may be considered, and lists aspects of each of these for consideration.

Construction kits

Construction kits are an important resource for design and technology activities, and we have selected one, Technical Lego, as an example of using such kits in the classroom.

Notes on the use of technical kits in the classroom

The suggestions in these notes can be used to assist teachers in the organization and management of technical construction kits in general, and Technical Lego in particular.

1 Pupils come to school with different previous experiences of construction kits. These experiences are very important in building confidence, since pupils have had time to 'tinker' and experiment. Pupils with little or no access at home need to be given greater opportunities to do so in school. There are many types of Lego sets on the market: Basic Lego (BL), Technical Lego without motors (TL1), and Lego sets with motors (TL2).
2 Teachers in a school need to decide how construction kits will be used in the classroom, as a choice 'play' activity during wet breaks and lunchtimes, or as part of the basic educational equipment in the classroom, in the same way as mathematics or science equipment. Any possible gender associations can be tackled by treating the use of construction kits as an educational rather than an optional 'play' activity.

	Clays & Plasticine	Fabrics/ Textiles	Food	Metal	Paper Card	Plaster Putties	Plastics	Rope & String	Waste Materials	Wood
Classification										
Characteristics/ Properties										
Sources/Production										
Fabrication methods: ● Cutting										
● Moulding/ Shaping										
● Joining: Temporary Permanent										
● Special techniques										
Adhesives										
Finishes										
Environmental implications										
Uses in history and across the world										

Figure 17 Aspects of different materials: a sample chart

3 Working in groups of twos with the Lego sets is a very useful activity for language development. It can be used to encourage team effort and co-operation. Teachers need to consider these groupings very carefully. This will be partly dependent on the class dynamics. Some teachers have found it useful to start with friendship groups to ensure greater chance of equal 'hands-on' experience. Mixing the groups at a later stage is equally important, to encourage pupils to learn to share and think about how they work together. Monitoring these group interactions is important, so that the teacher can observe if one pupil is constantly handling the Lego whilst the other is observing and can then suggest a change of roles.
4 Many construction kits come with workcards showing pupils how to build models. The Lego workcards show the principles of mechanical movements with examples of everyday applications. Some teachers find these a very useful introduction to the kits. However, it is important to note that pupils should be encouraged to use the construction kits as a tool to solve their own technical problems.
5 Constructing pupils' own models can be used as an excellent vehicle for observational drawing work, to encourage pupils to keep a record of what they have built. A drawing is one way of doing this. A useful suggestion is to ask each group to build their own model, produce instructional drawings for another group on how to build it, and pass it on to yet another group who will test their instructions before attempting their own.

Ideas for projects using Lego 2 or any other technical construction kit with a motor

The following problems can be incorporated into a variety of topics. The first problem is to construct the slowest moving motorized transporter. This problem enables the user to understand the concepts of gears and pulleys and their use for changing speeds. All the problems can also be adapted to computer control projects.

1 A slow moving transporter. Pupils can attach lights, buzzers and sequence the movements.
2 Construct a crane mechanism that lifts and drops a weight as carefully as possible. This can be part of a number of machines on a factory system of conveyor belts, or part of a group of machines.
3 Design and build a sliding door similar to those found in supermarkets. A series of sensors can be used to activate the doors and count the number of customers entering.
4 Build a motorized buggy that can go left and right as well as forwards and backwards. The class can sequence their own turtle.

5 Design and build a lift system. This can respond to calls from the different floors.
6 A barrier similar to those found in car parks or at railway junctions. These will be activated by the presence of a car. Counting the maximum number of cars in the car park and indicating whether the car park is full or empty. Otherwise it can be part of a railway system, including the train itself and a series of signals.
7 Different methods of lifting bridges to allow boats through.
8 A ski lift. This can be extended to allow skiers to get off before it moves on. It can be built as individual seats or as carriages to hold a number of skiers.
9 Motorized water-well.
10 Motorized portcullis or garage door.
11 A rotating display for a precious Fabergé egg (Easter egg is a good model). Pupils will have to design a system to protect the egg while it is on public display.
12 A robot arm. Many different actions and movements occur in these. Choose one and build it.

Activity sheets

Experience of constructing mechanisms can be introduced through paper engineering, and Figure 18 shows an activity sheet which addresses the concepts and processes involved. Activity sheets for introducing paper engineering, electricity, and pneumatics and hydraulics are shown in Figures 18, 19 and 20.

Curriculum development in design and technology: writing a school policy

In chapter 7 we addressed whole school development, giving as examples case-studies of two schools, where, with different focuses, design and technology activities were introduced and whole school policies developed. In this final section of appendix 2 we have outlined suggestions for forming a whole school policy for design and technology. We have included a possible methodology based on the experiences of the schools we have worked with. We also offer points for consideration when discussing and writing up the policy document. Finally we pose some questions to help address the relationship of design and technology to other areas of the curriculum.

Methodology

1 What is the focus of design and technology curriculum development in our school?

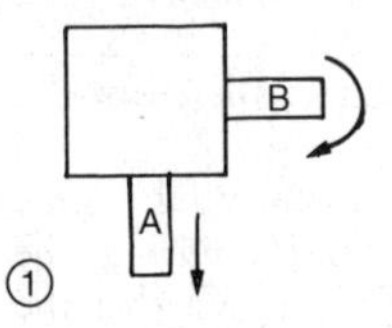

When you pull down lever A, lever B rotates. Are there any ways of controlling the extent of the movements of lever B?

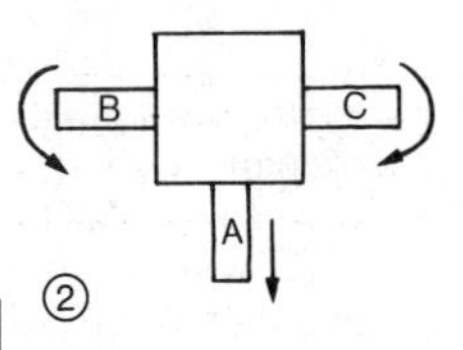

Extend idea 1 so that levers B and C rotate as lever A is pulled up and down. Can you think of 'easing' the movements of B and C?

SOME CONCEPTS
• AREAS OF UNDERSTANDING
• TECHNICAL TERMS
that can arise from this work.

MECHANICAL MOVEMENT
CAUSE AND EFFECT
ACCURACY
PRECISION IN MAKING
LEVERS
PIVOTS
CAMS
FOLLOWERS
CONNECTORS
GUIDES
FIXED POINTS
MOVING POINTS
POSITION OF PIVOTS
DIRECTION OF MOVEMENT

SOME PROCESSES & SKILLS INVOLVED IN THIS WORK

TECHNICAL
PROBLEM-SOLVING
HYPOTHESIZING
TRYING OUT
GENERAL TESTING
FAIR TESTING
IMPROVING SOLUTIONS
APPLYING TO MORE
OPEN-ENDED PROBLEMS
SUBJECTIVE AND
OBJECTIVE EVALUATIONS
CONSTRUCTING
MEASURING

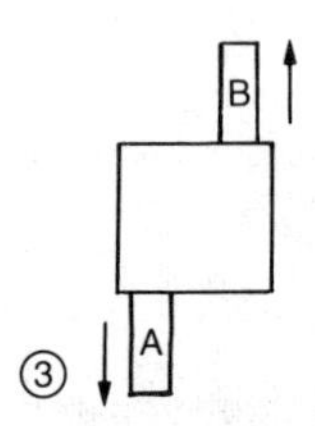

When you pull down lever A, lever B goes up and vice versa.

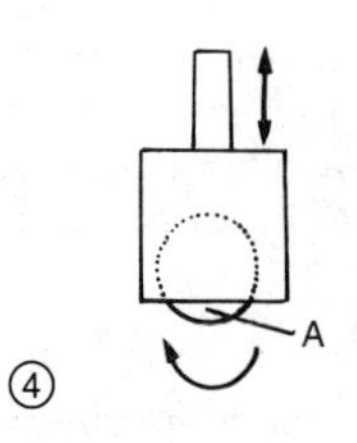

When you turn disc A, lever B goes up and down.

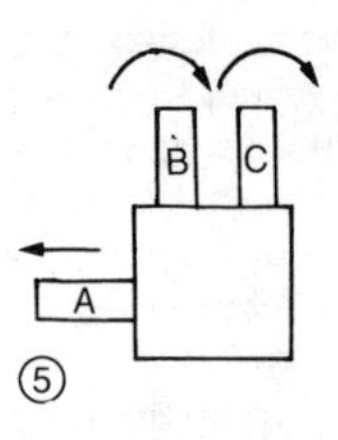

This is an extension of idea 1. When you pull lever A, lever B and C rotate in the same direction.

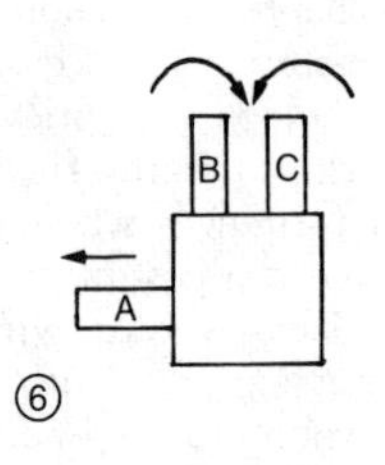

An extension of idea 5. When you pull lever A, levers B and C rotate in opposite directions.

Figure 18 Activity sheet: starting points for mechanical paper engineering projects

These are some activities and questions you may wish to address that would raise some concepts and technical terms that are relevant to the area.

1 Try lighting a bulb.
2 Try making the buzzer sound.
3 Try making the motor go round.
4 Can you make the motor reverse?
5 Make a break in your circuit. Try out different materials across the break to see if they will allow the electrical current through.
6 Can you make your own switch?
7 Can you make your own bulb holder?
8 Try lighting two bulbs. Do you notice anything? Are they both as bright as when you lit the bulb in 1?
9 Can you find the symbols for a light bulb, wires, your power source and your switch?
10 How is electricity generated? How could you generate a small electrical current from a waterwheel?

- **Concepts involved**
- **Areas of understanding**
- **Technical terms**

What is a circuit?

Power source

Load

Conductors

Polarity/reversing polarity

What is electricity?

Electrical current

Conductive materials

Non-conductive materials

Semiconductive materials

In series

In parallel

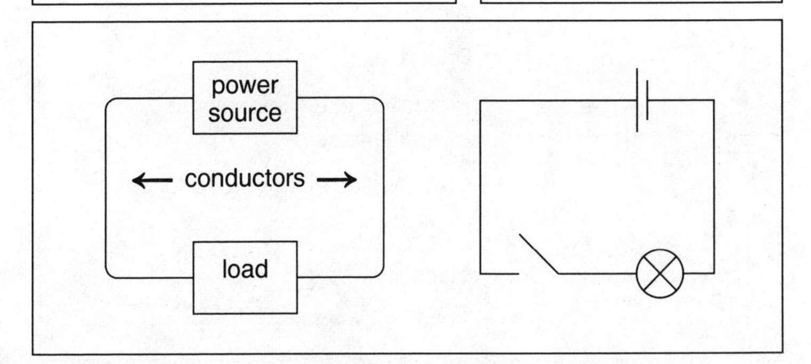

Figure 19 Activity sheet: introduction to electrical circuits

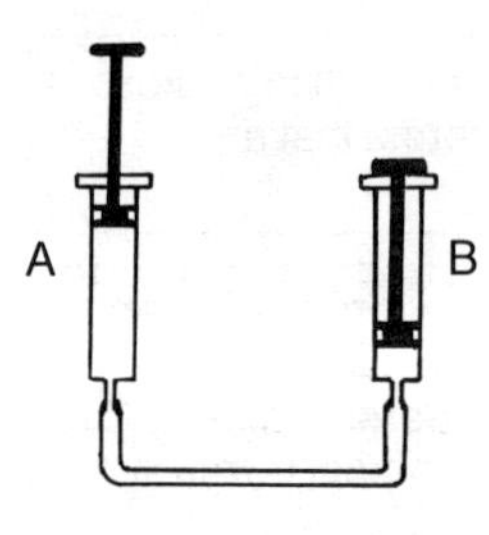

Diagram 1

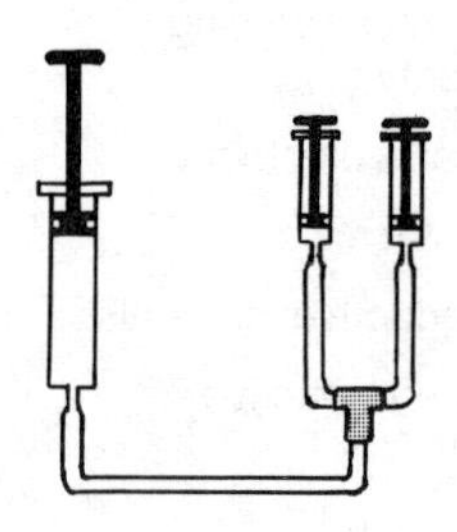

Diagram 2

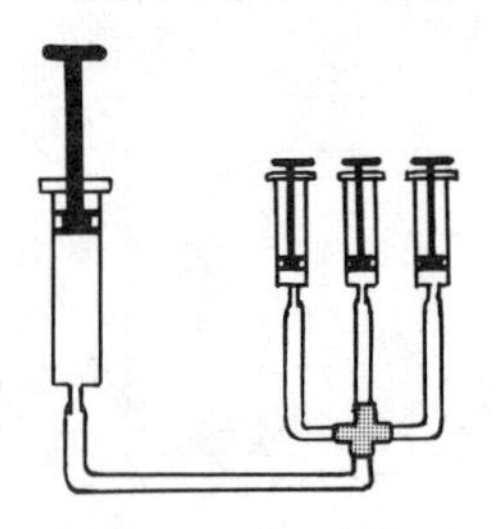

Diagram 3

1 What do you think would happen if:

a Plunger A is pushed down? Why?

b Plunger A is pulled up again?

c Syringe B is smaller than syringe A?

2 How much weight could a syringe lift?

3 Try filling the system with water. Avoid any air bubbles. Do you notice any differences in how the plungers work? Try the above tests with the system filled with water.

4 Now try with oil. What differences do you notice?

5 Try using a T-connector piece and connecting two syringes – see sketch.

6 Try using a crossover connector that would allow you to use three syringes – see sketch.

Areas of Understanding – Concepts raised

Air pressure

Air compression

Pneumatics

Properties of air: sponginess: springiness

Compressed air is a powerful force

Hydraulics

Viscosity

Relative air volume

Figure 20 Activity sheet: pneumatics and hydraulics

Examples:
- in relation to topic work
- in relation to problem solving
- in the National Curriculum
- in relation to language, science, mathematics, aesthetic and creative areas
- Equal opportunities

Discussion and agreement among staff as to the needs of in-service training and development and the focus of it within the school.

2 Design a general plan of action for school development.
- staff attending in-service
- LEA inspectors and advisory teachers support and advice
- visiting other schools
- initial resources and materials
- responsibility for overseeing work
- responsibility for recording and monitoring overall process (inset co-ordinator?)
- programme of staff workshops
- programme of staff discussions on issues arising
- programme for implementation in the classrooms

3 Period of implementing the above programmes and structures.
- monitoring and recording the process
- periodic reviews of the work
- discussions on emerging issues (see below)

4 Review process, involving discussion on the initial implementation, recording work in classrooms, reflecting on the work.
- dissemination of successful applications
- review of problematic areas
- discussions on issues arising

5 Writing a school policy based on the above experiences.
6 What sort of policy do we want to write, what is its function? Who is it for? How will it be reviewed?
7 How will new members of staff be introduced to the practicalities of the school policy?

Points for consideration in a school policy for design and technology

1 Where do design and technology activities fit in the primary curriculum? Where do they fit into our school policies?
2 What do we mean by design and technology? How do we define the subject? Can we list all the types of design and technology activities that teachers have been involved in.
What materials did we need for these? How did we approach them?
3 What is their relationship to topic or theme work or project work?

General and specific comments. What decisions do we wish to make relating to this?

4 How do we incorporate design and technology activities in our work, monitor their implementation, and record progress in terms of the National Curriculum?

5 What do we mean by problem solving? What are our aims in regard to developing problem-solving activities throughout the school?

6 What documents, booklets, reports have we found useful and pertinent to this policy (so that a newcomer may refer to these)?
Are there any we feel every member of staff should have?

7 Are there any specific recommendations/approaches that we have decided upon in the light of the above? Are there particular activities or methods of working that we wish to encourage?

8 How do we ensure that equal opportunity policies are implemented with regard to the use and access to tools, materials, construction kits, as well as types of projects, classroom grouping, resources (books, posters, visual aids, videos), both in terms of gender and culture?

Other areas of consideration include skills, techniques and useful resources. The process of design and make activities in schools has meant the acquisition of certain skills and techniques by members of staff on INSET courses. Is it worth recording these to have a shared knowledge that can be continuously built upon?

- paper engineering
- specific methods of construction
- three-dimensional modelling techniques
- use of tools (which tools? how and where are they stored? who is responsible? what about safety recommendations? which tools are shared and which remain in classrooms, are there any issues of security?)
- wheels, pulleys, gears (construction, purpose and function)
- electrical circuits
- use of materials
- principles of mechanical movement
- motorizing
- energy (forms of, topics related to concepts involved)
- computer control
- construction kits (how do we use them, store them, how do they fit into general work, how can we relate to other materials, advantages and possible disadvantages of working with kits? see notes and ideas for Technical Lego)

Relationship to other areas of the curriculum

1 Science. What common methods and processes exist? How can design and technology activities enhance scientific processes and knowledge?
2 Language development. List all the different language activities that arise from this form of work as has been carried out in the past. Can the staff make any recommendations with regard to these and the development of a school policy?
3 Mathematics. Similar to language and science and technology: list all the mathematical activities/investigations/concepts engaged in throughout the projects. Can the school make any recommendations that will relate to links with mathematics that can be incorporated into an agreed school policy?
4 Social studies. What context was the work put into? Does this give a historical and/or geographic perspective? Are pupils aware of the social issues and values that underlie the work? List projects and see how they link or can be adapted to include these links. How can we devise situations that give rise to more open-ended problems that involve pupils in handling these concepts?
5 Environmental studies.
6 Aesthetic and creative. What skills, techniques and materials can be introduced and incorporated into a design and technology policy so that we are not repeating unnecessarily, and are maintaining continuity? And which of these require specific consideration separately?
7 Information technology.

References

Department of Education and Science (1989a) *English in the National Curriculum*, London: HMSO.
—— (1989b) *Mathematics in the National Curriculum*, London: HMSO.
—— (1989c) *Science in the National Curriculum*, London: HMSO.
—— (1990) *Technology in the National Curriculum*, London: HMSO.
Design Council Report (1987) *Design and Primary Education*, London: Design Council.
Fisher, S. and Hicks, D. (1985) *World Studies 8–13*, London: Oliver & Boyd.
Flood, J. (1986) *Tools for Technology*, Ely Resource and Technology Centre.
Grant, M. and Harding, J. (1984) *Presenting Design and Technology to Girls* (*Girls and Technology Education Report 84:2*), London: Chelsea College.
H.M.I. (1987a) *Craft, Design and Technology 5–16* (*Curriculum Matters 9*), London: HMSO.
—— (1987b) *The Curriculum from 5 to 16*, 6th imp., London: HMSO.
McInnes, Ovetta (1987) *Recommendations for Teachers*, Haringey: Multi Cultural Support Group.
Myers, K. (ed.) (1987) *Gender Watch: Self-assessment Schedules for use in Schools*, London: School Curriculum Development Committee Publication.
National Curriculum Council (1990) *Design and Technology: Non-statutory Guidance for Design and Technology*, London: HMSO.
National Curriculum Design and Technology Working Group 1988 *The Design and Technology Interim Report*, London: HMSO.
—— (1989) *Design and Technology for Ages 5–16: Proposals for the Secretary of State for Education and the Secretary of State for Wales.*
Polyani, M. (1973) *Personal Knowledge: Towards a Post-critical Philosophy*, 3rd repr., London: Routledge & Kegan Paul.
Shallcross, P. (1985) *Starting Technology: The Simple Approach*, London: Arnold.
Sparkes, J. (1987) 'Technological Education', in *Living with Technology: A Course for Teachers*, Milton Keynes: Open University Press.

Williams, P. and Jinks, D. (1985) *Design and Technology 5–12*, London: Falmer Press.

Index